A TEXT BOOK OF

PHYSICS

Solid State Physics
(Paper - XVI)

FOR

B.Sc. Part - III : Semester - VI

As Per New Revised Syllabus of Shivaji University, Kolhapur, June 2015

Dr. M. G. PATIL
M. Sc., M. Phil, Ph. D.
Associate Professor and Head
Department of Physics,
D.K. A.S.C. College, Ichalkaranji

Dr. R. H. PATIL
M. Sc., Ph. D., NET (L & JRF)
Department of Physics
Smt. K. R. P. Kanya Mahavidyalaya,
(Arts, Comm. and Science), Islampur

Dr. L. D. KADAM
M. Sc., Ph. D.
Associate Professor,
Department of Physics,
Y. C. Institute of Science, Satara

S. V. MALGAONKAR
M. Sc.
Associate Professor,
Department of Physics,
Vivekanand College, Kolhapur

N1228

B.Sc. Part - III : Physics (P-XVI) (Sem. VI)   **ISBN 978-93-5164-872-7**

Fourth Edition   : **January 2019**

©   : **Authors**

Published By:

NIRALI PRAKASHAN

Abhyudaya Pragati, 1312, Shivaji Nagar

Off J.M. Road, PUNE – 411005

Tel - (020) 25512336/37/39, Fax - (020) 25511379

Email : niralipune@pragationline.com

➤ DISTRIBUTION CENTRES

PUNE

Nirali Prakashan : 119, Budhwar Peth, Jogeshwari Mandir Lane, Pune 411002, Maharashtra
(For orders within Pune)   Tel : (020) 2445 2044, 66022708, Fax : (020) 2445 1538; Mobile : 9657703145
Email : niralilocal@pragationline.com

Nirali Prakashan : S. No. 28/27, Dhayari, Near Asian College Pune 411041
(For orders outside Pune)   Tel : (020) 24690204 Fax : (020) 24690316; Mobile : 9657703143
Email : bookorder@pragationline.com

MUMBAI

Nirali Prakashan : 385, S.V.P. Road, Rasdhara Co-op. Hsg. Society Ltd.,
Girgaum, Mumbai 400004, Maharashtra; Mobile : 9320129587
Tel : (022) 2385 6339 / 2386 9976, Fax : (022) 2386 9976
Email : niralimumbai@pragationline.com

➤ DISTRIBUTION BRANCHES

JALGAON

Nirali Prakashan : 34, V. V. Golani Market, Navi Peth, Jalgaon 425001, Maharashtra,
Tel : (0257) 222 0395, Mob : 94234 91860;
Email : niralijalgaon@pragationline.com

KOLHAPUR

Nirali Prakashan : New Mahadvar Road, Kedar Plaza, 1st Floor Opp. IDBI Bank, Kolhapur 416 012
Maharashtra. Mob : 9850046155; Email : niralikolhapur@pragationline.com

NAGPUR

Nirali Prakashan : Above Maratha Mandir, Shop No. 3, First Floor,
Rani Jhanshi Square, Sitabuldi, Nagpur 440012, Maharashtra
Tel : (0712) 254 7129; Email : niralinagpur@pragationline.com

DELHI

Nirali Prakashan : 4593/15, Basement, Agarwal Lane, Ansari Road, Daryaganj
Near Times of India Building, New Delhi 110002 Mob : 08505972553
Email : niralidelhi@pragationline.com

BENGALURU

Nirali Prakashan : Maitri Ground Floor, Jaya Apartments, No. 99, 6th Cross, 6th Main,
Malleswaram, Bengaluru 560003, Karnataka; Mob : 9449043034
Email : niralibangalore@pragationline.com

Other Branches : Hyderabad, Chennai

niralipune@pragationline.com | www.pragationline.com

Also find us on 🇫 **www.facebook.com/niralibooks**

PREFACE

This book is primarily intended for B.Sc. - III students of Semester - VI course of Shivaji University, Kolhapur. The book is strictly written according to new syllabus prescribed by Shivaji University, Kolhapur from June 2015.

It is our great pleasure to present this book to the students and respected teachers in proper time. The subject matter is presented in lucid and simple language. The book covers all the chapters according to the syllabus. The material is presented in comprehensive way and sequence of articles in each chapter helps the student to understand the subject. Different diagrams are given in the book to understand the basic principles. The solved numerical examples, multiple choice type questions, short answer type questions and long answer type questions are given at the end of each chapter. Some numerical problems are also given for self study.

We are heartly thankful to Principal Abhaykumar Salunkhe, President, Mrs. Shubhangi Gavade, the Secretary and Dr. Ashok Karande, Joint Secretary, Swami Vivekanand Shikshan Sanstha, Kolhapur, Dr. Arvind Burangale, Secretary, Rayat Shikshan Sanstha, Satara, Dr. H. B. Patil, Principal, Vivekanand College, Kolhapur. Dr. Milind Hujare, Principal. D.K.A.S.C. College Ichalkaranji, Dr. S. V. Kakatkar and Dr. V. C. Mahajan, Vivekanand College, Kolhapur, who inspired us to write this book.

One of the authors Dr. R. H. Patil is thankful to his daughter Vaishnavi, son Rajvardhan, wife Suvarna and the mother Parvati. Also the author Dr. M. G. Patil is thankful to his wife Mangal, sons Abhijeet and Yuvraj, grand daughter Aaradhya and the mother Hirabai.

We are thankful to Nirali Prakashan, Pune for making us a part of their team of Authors. We thank Mr. **Dineshbhai Furia** and **Mr. Jignesh Furia** for publishing this book.

We also thank Mr. Girish Redkar (Heak Marketing Dept.) for his co-operation in publishing this book.

Last but not the least we are very much indebted to Mr. Virdhaval Shinde, (Marketing Executives, Kolhapur District) and Mr. Ashok Nanavare (Marketing Executive, Sangli District) for their nice co-operation. We are very much thankful to Mr. Kiran Velankar (Proof Reading), Ms. Chaitali Takle and Mr. Santosh Bare for a neat and error free D.T.P. of this book.

We hope that this book will be found useful to the students and teachers. We will appreciate any suggestions for the improvement of the book.

– **Authors**

SYLLABUS

Unit - I (11)

Crystal Structure

Crystalline and non-crystalline solids, Space lattice, Basis and crystal structure, Unit cell – Primitive and non-primitive, Bravais lattices – Space groups and crystal structures, Symmetry elements of cubic system, Miller indices, Relation between lattice constant, Interplaner spacing and Miller indices, Simple crystal structures – Cubic (SC, BCC, FCC) and hexagonal close packed (HCP) (with respect to coordination number, Atomic radius, Atoms per unit cell, Packing fraction).

Unit - II (11)

X-Ray Diffraction by Crystals :

Reciprocal lattice, Properties of reciprocal lattice, Bragg's law in reciprocal lattice (Ewald's construction), Powder method of X-ray diffraction and analysis of cubic crystal structure.

Unit - III (11)

Lattice Vibrations :

Elastic vibrations of linear one-dimensional monoatomic lattice, Expression for frequency and dispersion curve, Elastic vibrations of linear one dimensional diatomic lattice – optical and acoustical excitations in ionic crystals, Experimental determination of dispersion relations.

Unit - IV (12)

1. Free Electron Theory of Metals and Band Theory of Solids :

Sommerfield's free electron model for electrical conductivity of metals, Fermi-Dirac distribution, Origin of energy bands - Valence band, Conduction band, Band gap energy, Distinction between metals, semiconductors and insulators, Hall effect – Hall voltage and Hall coefficient.

2. Solid State Device :

Timer (IC 555) - Block diagram, Function of each block, Pin configuration, Applications – Astable, Monostable, Bistable multivibrator.

❑❑❑

CONTENTS

❑❑❑

1

CHAPTER

CRYSTAL STRUCTURE

SYLLABUS

Crystalline and non-crystalline solids, Space lattice, Basis and crystal structure, Unit cell – Primitive and non-primitive, Bravais lattices – Space groups and crystal structures, Symmetry elements of cubic system, Miller indices, Relation between lattice constant, Interplaner spacing and Miller indices, Simple crystal structures – Cubic (SC, BCC, FCC) and hexagonal close packed (HCP) (with respect to coordination number, Atomic radius, Atoms per unit cell, Packing fraction).

1.1 INTRODUCTION

The matter is usually regarded to exist in solid state or fluid state. Fluid state is further divided into gaseous and liquid state. All the materials are composed of atoms and molecules. The solid state of matter can be put into two categories on the basis of their structure i.e. crystalline solid and non-crystalline solid (amorphous). They are distinguished from one another primarily by the degree of order exhibited by the arrangement of the fundamental particles – atoms, molecules or ions comprising them.

Crystallography is the study of the solid in the form of crystals. The discovery of X-rays gives the powerful and precise tool for study of the internal arrangement of atoms in the crystal. Once the internal arrangement of atoms in the crystal is known, their physical properties can be studied.

1.2 CRYSTALLINE AND NON-CRYSTALLINE SOLID

In crystalline solid, the atoms or the molecules or ions are stacked in a regular (or periodic) manner, thus forming a three dimensional pattern. Smallest group of atoms called pattern unit or building block, which repeat itself in all directions to form a crystal.

When the regularity of the pattern extends throughout a certain piece of solid, then it is called single crystal. However, most of the solids are not single crystals, but often consist of a large number of small single crystal sections (grains) of various shapes and sizes packed to one another along the interfaces called the grain boundaries. In such materials the regularity or periodicity is interrupted at the grain boundaries, these materials are called polycrystalline.

When the size of grain becomes comparable to the size of pattern unit, the periodicity structure is completely disturbed, then the material is said to be in an amorphous state.

Distinction between Crystalline and Non-crystalline Solid :

Crystalline solid	Amorphous solid
1. Regular arrangement of particles or atoms.	1. Random arrangement of particles or atoms.
2. They have different physical properties (thermal, electrical) in different directions, i.e. they are anisotropic.	2. They have same physical properties in all directions i.e. they are isotropic.
3. The cooling curve for them has breaks because of crystallization.	3. The cooling curve is smooth.
4. The melting point is very sharp.	4. The melting point is not sharp.
5. Due to slow growth process, the constituent particles take definite position, where the potential energy of the configuration is minimum during growth. A long range order exists in crystalline solid.	5. Due to the quick growth process or phase change, the atoms do not have sufficient time to obtain the configuration of minimum energy. Therefore, short range order occurs in amorphous solid. e.g. glass, plastics, etc.

The crystals are formed when a substance changes from one state to another state i.e. in phase transformation. A change from the liquid phase to solid crystallization and a change from gaseous phase to the solid crystallization occur by sublimation. Once the crystal is formed, the atoms or molecules (building blocks) are bonded by chemical bonds like covalent bond, ionic bond, metallic bond and Van der Waal's bonds, etc. The crystals are bounded by perfectly flat faces and have sharp and straight edges. Crystal exhibits certain symmetries. Now-a-days X-rays, electron beam, neutron beam provide a tool for studying the internal structure of crystal. The study of the geometric form and other physical properties of crystalline solid by using X-ray diffraction, electron diffraction or neutron diffraction is called science of crystallography.

Different features of crystals :

1.   Faces : The crystals are bounded by number of perfectly flat surfaces. These surfaces are called as faces. The face of the crystal may be like (cubes, alum) or unlike (galena).

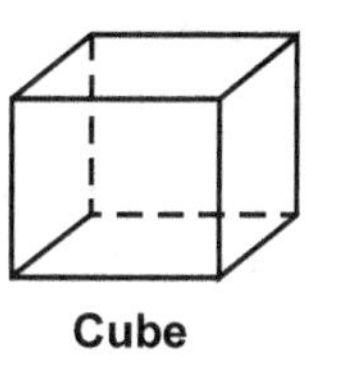
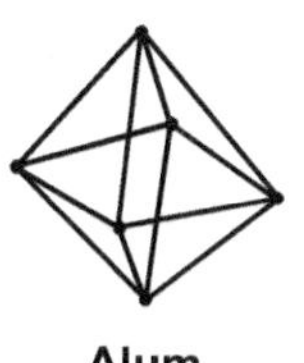

Cube   **Alum**   **Galena**

Fig. 1.1

2.   Form : All the faces corresponding to a crystal are said to constitute a form. The crystal that consist of all like faces is termed to have a simple form, while the crystal having two or more simple forms is called to have a combination form.

3.   Edges and interfacial angles : The intersection of two adjacent faces form the edges of crystal. The angle between any two faces of crystal is termed as the interfacial angle. The relation between plane faces (f), straight edges (e) and interfacial angle (c) is given by

$$f + c = e + z$$

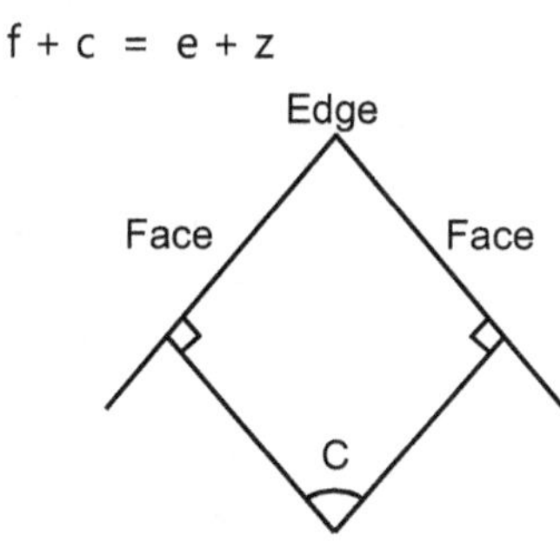

Fig. 1.2

1.3 SPACE LATTICE

Crystals are made up of regular and periodic three dimensional pattern of atoms in space. From geometrical point of view every atom is repetition of some arbitrarily chosen 'motif' atom of the mass. When a motif is repeated systematically in three dimensions, it results in a pattern called space lattice. Thus space lattice is defined as the periodic array of points in space such that the environment about each point is the same. In other words, a lattice is a periodic arrangement of points in space. The environment about any other point is in every way the same as that about any other point. All the points are connected by the translation operation T which is defined as $\vec{T} = n_1 \vec{a} + n_2 \vec{b} + n_3 \vec{c}$, where n_1, n_2 and n_3 are arbitrary integers and $\vec{a}$, $\vec{b}$ and $\vec{c}$ are fundamental translation vectors.

If the array of points form a plane, it is two dimensional lattice and if the array of point repeats in three dimensions, it is three dimensional space lattice or simply space lattice.

(a) Two dimensional lattice   **(b) Two dimensional collection of points (not lattice)**

Fig. 1.3

In Fig. 1.3 (a) and (b), array of point in two dimensions are shown. In Fig. 1.3 (a) we observe that the environment about any point is same but in Fig. 1.3 (b) it is not the same.

1.4 BASIS AND CRYSTAL STRUCTURE

Since a point being an infinitesimal spot in space, is imaginary, a lattice of point is an imaginary concept because a point is a dimensionless quantity. A lattice is a mathematical concept whereas the crystal structure is a physical structure. Crystal is formed by associating identically with every lattice point, a structural or building unit. This structural unit is called the basis or pattern. This basis consists of an atom or group of atoms. When the basis is repeated with correct periodicity in all direction it gives the actual crystal structure. (The lattice points may or may not be the atom sites.)

The logical relation between crystal and lattice is

Lattice + Basis = Crystal structure

The distinction between crystal and lattice is illustrated in Fig. 1.4.

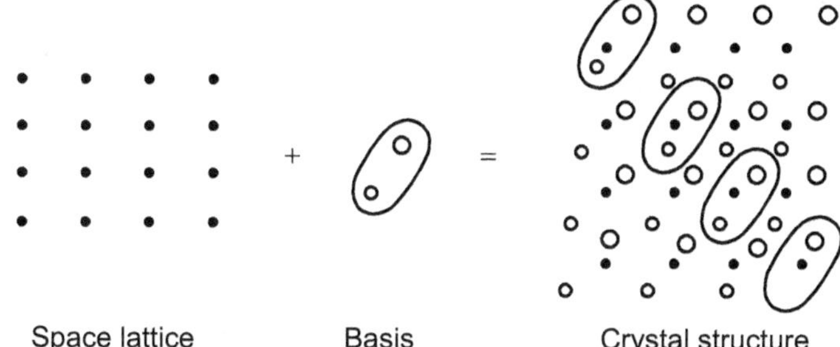

Space lattice   Basis   Crystal structure
Fig. 1.4 : Crystal structure and basis

Thus crystal structure is specified characteristically by the type of the lattice associated with it.

1.5 UNIT CELL – PRIMITIVE AND NON-PRIMITIVE

An ideal crystal is constructed by a regular repetition in space of identical structural unit or building blocks usually termed as unit cell of the crystal. Thus unit cell of the crystal is the fundamental unit, like a brick (which may consist of a group of atoms, molecules or ions) which repeat at regular interval in three dimensions to form a crystalline solid.

For the very simplest of the monoatomic crystals, the building block is a single atom, but in most crystals the building block contains several atoms or molecules.

It is usually to choose as a unit cell, a parallelopiped formed by lattice points only at their corners. If unit cell contains only one lattice point, it is called primitive (simple) cell. On the other hand if a unit cell contains more than one lattice point, it is called non-primitive or multiple cell.

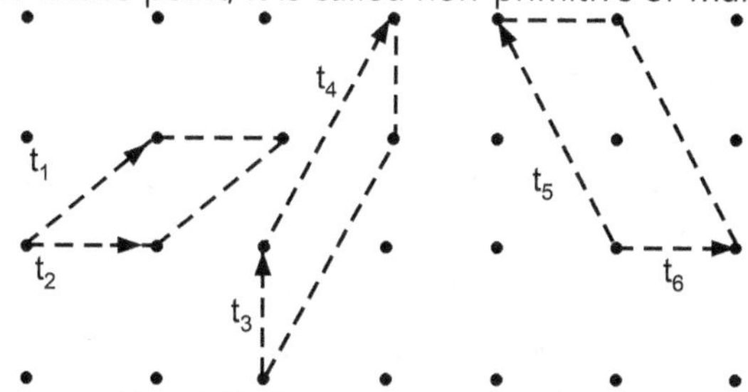

Fig. 1.5 : Formation of unit cell

In Fig. 1.5, if pairs of translation such as $t_1 t_2$ or $t_3 t_4$ are chosen, they are said to definite a primitive cell because it contain only one lattice point (or atom). If translation like t_5, t_6 are chosen, the unit cell contain more than one lattice point and it is non-primitive cell.

Number of lattice points per unit cell :

In two dimensional space lattice, each unit cell is considered as a parallelogram which has a representative lattice point at its each corner. Each corner lattice point is common to four similar unit cells and its contribution to a particular cell is $\frac{1}{4}$. Since there are four corners to unit cell, the total number of lattice points per unit cell will be $\frac{1}{4} \times 4 = 1$. Such cell is called the primitive cell.

In three dimensional space, an elementary rectangular parallelopiped (mutually non-coplaner translation) represents a most general unit cell. A rectangular parallelopiped has a lattice point at its corner. Also each corner is common to eight similar unit cells. Therefore its contribution to a particular cell is $\frac{1}{8}$. There are eight corners to the unit cell. Therefore total number of lattice points per unit cell is $\frac{1}{8} \times 8 = 1$.

Thus both in two dimension and three dimension space lattice, each cell contain only one whole lattice point.

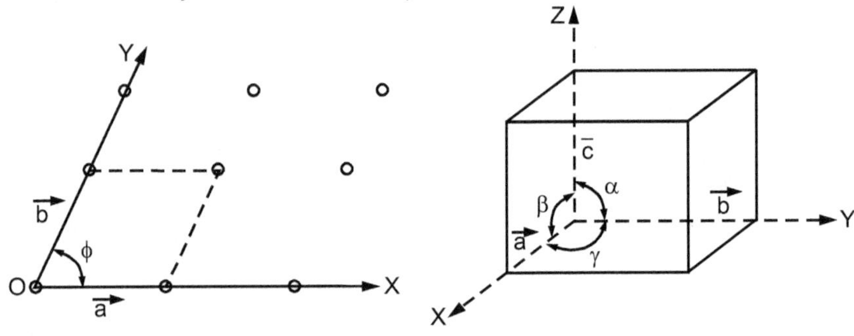

(a) Two dimensional space lattice unit cell

(b) Three dimensional space lattice unit cell

Fig. 1.6 : Lattice parameters

The unit cell is defined by the length of edges and the angles between them are called lattice parameters. In two dimensional space lattice, unit cell [Fig. 1.6 (a)] is represented by the vectors $\vec{a}$ and $\vec{b}$ and angle ϕ between them. In three dimension space lattice, unit cell [Fig. 1.6 (b)] is represented by vectors $\vec{a}$, $\vec{b}$ and $\vec{c}$ as length of edges (sides) which are called crystallographic axes and the α, β and γ as the angles between corresponding crystal axes. The volume of unit cell is

$$V = \left| (\vec{a} \times \vec{b}) \cdot \vec{c} \right|.$$

1.6 SYMMETRY ELEMENTS OF CUBIC SYSTEM

A symmetry operation (geometrical operation) is a transformation performed on the body (crystal) which leaves it unchanged or invariant i.e. if a body attain its initial configuration even after performing operation on it then the body is said to possess a symmetry corresponding to that operation. In other words, if the environment of crystal (or cubic system) remain unchanged after performing symmetry operation on it then the crystal (cubic system) is said to possess a symmetry corresponding to that operation.

There are four types of symmetry elements for cubic system i.e. there are four principal means for repeating motif in space.

I. Translation :

Translation by an amount T keep the crystal environment invariant.

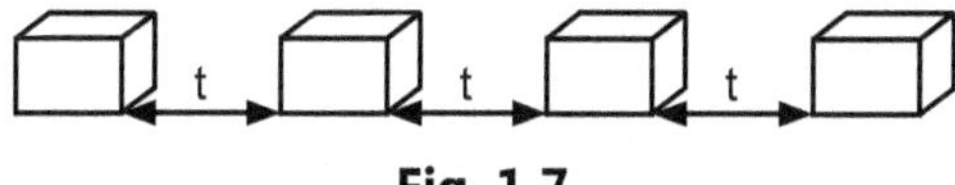

Fig. 1.7

II. Rotational axis of symmetry (n-fold axis of symmetry) :

A crystal is said to possess rotational symmetry about an axis if the rotation of the crystal about this axis by some angle θ leaves the crystal invariant.

A cube is said to possess n fold rotation axis of symmetry if rotation through $\left(\dfrac{360}{n}\right)$ about an axis brings the cube into self coincidence (congruent position), where the integer n is called the multiplicity of the rotation axis. n can have only the value 1 (single fold axis), 2 (diad axis), 3 (triad axis), 4 (tetrad axis) or 6 (hexad axis) i.e. only five rotation axis in all.

1. If n = 1, the cube must be rotated through $360°$ to achieve congruence. Such axis is called identity axis and every crystal possesses an infinite number of such axes.

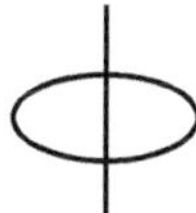

Fig. 1.8

2. If n = 2, the cube must be rotated through 180° to achieve congruence and the axis is called diad axis (⬭), each passing through the middle point of a pair of opposite parallel edges. Since there are twelve edges in the cube, pair will be six i.e. six axes of two fold symmetry.

3. If n = 3, the corresponding angle of rotation is 120° and the axis is called triad axis (▲), about each diagonal of the cube i.e. four axes of three fold symmetry.

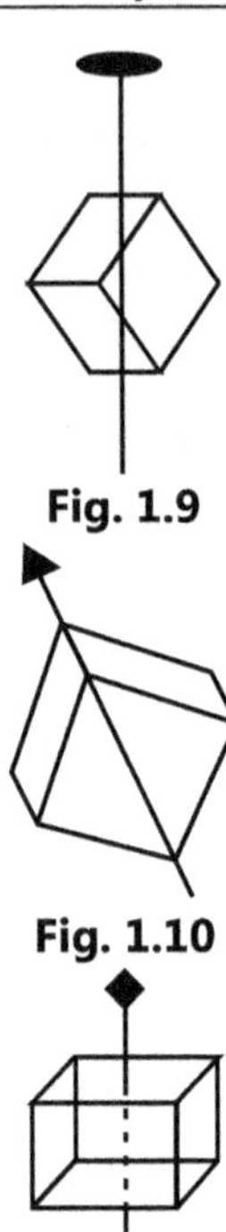

Fig. 1.9

Fig. 1.10

4. If n = 4, the angle of rotation is 90° and the axis of rotation is called tetrad axis (◆), one normal to each three pairs of parallel faces of cube i.e. three axis of four fold symmetry.

Fig. 1.11

5. If n = 6, the angle of rotation is 60° and the axis of rotation is called hexad axis (⬭).

Crystals do not show five fold or any other symmetry axis higher than six, because the crystal is not just a solid body, but identical atomic or molecular arrangement is periodic in three dimensions. Crystal has translation periodicity and identical repetition of a unit can take place only when we consider 1, 2, 3, 4 and 6 fold axes. Thus the pentagon lattice is not possible. Pentagon cannot be made to meet at a point, being a constant angle to one another.

Thus cube has 6 diad axes, 4 triad axes and 3 tetrad axes, the total rotational axis of symmetry for cube are 13.

III. Plane of symmetry (m) : Reflection planes or Mirror planes :

A plane can be drawn in the crystal which contains the centre of the crystal. Thus one half of the crystal is the reflection of the other half. Then the crystal is said to have a plane of symmetry. Thus if a crystal is cut along the plane and put it on a mirror, then the image will produce the other half of the crystal. Such a plane is known as plane of symmetry or symmetry plane and is represented by m.

In a cube, there are 3 planes of symmetry parallel to the faces of the cube. Also cube has 6 diagonal planes of symmetry from due to possible edges. Thus cube has in all 9 (i.e. 3 to 6) planes of symmetry.

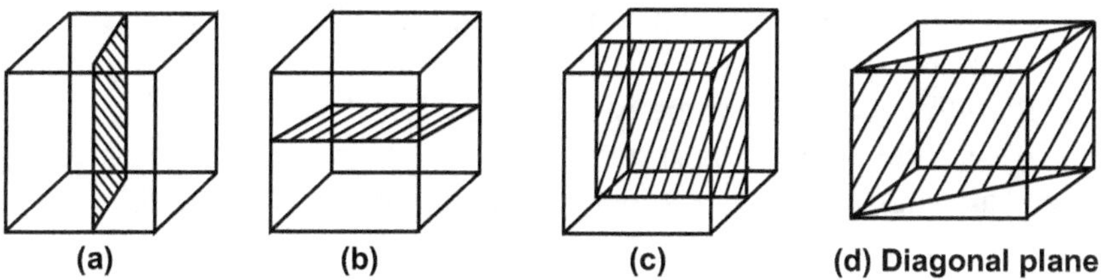

Fig. 1.12 : Planes of symmetry

IV. Centre of symmetry (Inversion centre) :

A crystal has an inversion centre, if the every lattice point given by a position vector $\vec{r}$, there is corresponding lattice point given by position vector $(-\vec{r})$.

In other words, the centre of symmetry is such a point in a crystal or body that if a line is drawn from any point on the crystal through the centre point and extended an equal distance on either side of this centre point, it will meet at an identical point. The inversion centre is denoted by i. The centre of symmetry leads to parallel pairs of faces on opposite sides of the crystal. The centre of symmetry represents reflection through a point instead of reflection of a plane.

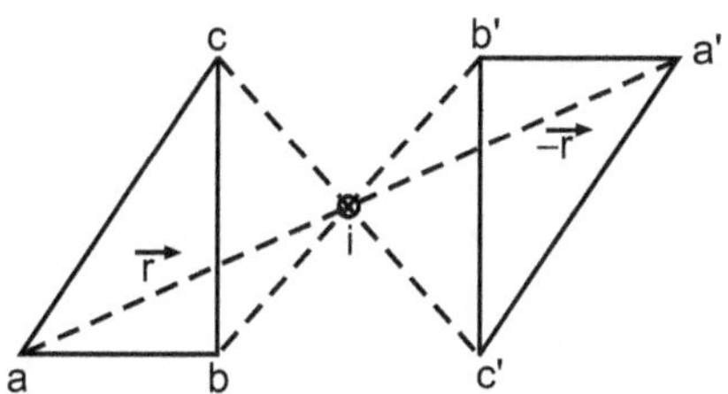

Fig. 1.13 : Centre of inversion (i)

Repetition by translation or rotation leaves the motif unchanged, whereas reflection or inversion changes the character of motif from right handed to left handed. In other words, translation or rotation symmetry produces congruent figures whereas reflection or inversion symmetries produces enantiomorphous (mirror image of each other) figures.

Thus the full crystallographic symmetry of a cube is :

In rotational axis of symmetry :

6 diads + 4 triads + 3 tetrads = 13 axes

In plane of symmetry : 3 planes + 6 diagonal planes = 9 planes and

In centre of symmetry = 1

Therefore, total 23 symmetry elements of a cube.

Translation repeat the motif, an infinite number of times along a given direction, while the other operation repeat the motif, a finite number of times.

V. Roto-reflection axis :

If a rotation is combined with a reflection into single hybrid operation, the resulting operation is called a roto-reflection. The corresponding symmetry element is called roto-reflection axis. There exists a roto-reflection axis corresponding to each of the five proper axis. Roto-reflection axis is represented by placing tilde ($\sim$) on numerical symbol of the corresponding axis.

VI. Roto-inversion axis :

Inversion is equivalent to reflection through a point. Rotation followed by inversion i.e. if inversion centre is combined with a rotation axis to produce roto-inversion axis, then crystal structure can be brought into self coincidence. Crystal can possess one, two, three, four and six fold roto-inversion axis.

1.7 BRAVIS LATTICE : SPACE GROUPS AND CRYSTAL STRUCTURE

Symmetry operation performed about a point or a line define point group symmetry elements and symmetry operations performed by a combination of rotation or reflection and translation both define space group symmetry elements.

Group means a set of elements which obey a certain set of rules. The symmetry elements that can exist in a crystal such as rotation, reflection and inversion operation also form symmetry group, point group and space group.

Point group : A point group in lattice is defined as the collection of symmetry operation which when applied about a lattice point leaves the lattice invariant. In other words, in point groups all the possible symmetry elements must pass through the point. Thus in a crystal each point has an associated point group. Therefore, there is an arrangement of point groups in space.

The various combination of allowed rotation and reflection operation gives rise to 32 allowed point groups. (Depending on the set of symmetry operations valid for a crystal, the crystals have been divided into 32 point groups.) Crystals belonging to different crystal systems show different point group symmetries and so the different crystal systems can be classified on the basis of point groups. Crystals belonging to the same point group have similar physical properties. (There are a total of ten possible point groups permissible in two dimensional crystal).

Space group : When point group symmetry operations are combined with translation symmetry element, space is obtained. Thus set of symmetry operation i.e. space group is obtained by combination of point group and translation symmetry. There are 230 space groups exhibited by the crystal. Such a large number of space group is (i) due to the addition of new king of symmetry elements (glide plane and screw axis) and (ii) because the planes and axes do not all pass through one point in a space group.

Glide plane : When a mirror plane is combined with a simultaneous translation operation we get a glide plane. The glide plane is always parallel to the plane mirror.

Screw axis : Rotation axis coupled with translation parallel to the rotation axis gives rise to this new symmetry element, the screw axis. As n screw fold axis produces a rotation by $\dfrac{2\pi}{n}$ about the axis and a translation parallel to the axis whose depends on n.

Characteristics of space group :

1. The symmetry of crystal structure is specified completely when the space group is known.

2. The space group through its symmetry elements determines the position of equivalent points within the unit cell.

3. The space group is characterized by Bravais lattice and by the location of the point group and other symmetry elements in a unit cell.

4. For a crystal structure specified by a particular space, if only one point in a cell is occupied, then it is necessary that all equivalent points will be occupied by identical atoms or molecules.

    (There are seventeen possible distinct space groups in two dimensions.)

Bravais Lattice :

Crystals occurring in nature reflect the symmetry of their internal atomic arrangements in the symmetry relating their terminal faces. Though of course there are deviation from ideal crystals at the surface where the periodicity condition obviously break down and in the interior due to dislocations, impurities and other imperfection, yet still the symmetry can be used to classify crystals.

To describe the structure of crystals, Bravais in 1848 introduced the concept of the space lattice. There are various ways of positioning structureless points in space such that all points have identical surroundings. These are known as Bravais lattice. There is no natural restriction on the length of translation vectors ($\vec{a}$, $\vec{b}$ and $\vec{c}$) and on the angle (α, β and γ) between them, an unlimited number of lattice is possible. But the lattice should be invariant under symmetry operations. There are five Bravais lattices in two dimensions and fourteen Bravais lattices in three dimensions.

(A) Bravais lattices in two dimensions :

There are five Bravais lattices in two dimensions, one general (oblique) and four special cases. The oblique lattice is invariant only under rotation of $\dfrac{2\pi}{1}, \dfrac{2\pi}{2}$ or under mirror reflection. But the special lattices of oblique type can be invariant under rotation $\dfrac{2\pi}{3}, \dfrac{2\pi}{4}, \dfrac{2\pi}{6}$ or under mirror reflection.

(a) Oblique lattice : $|\vec{a}| \neq |\vec{b}|$; $\phi = 90°$

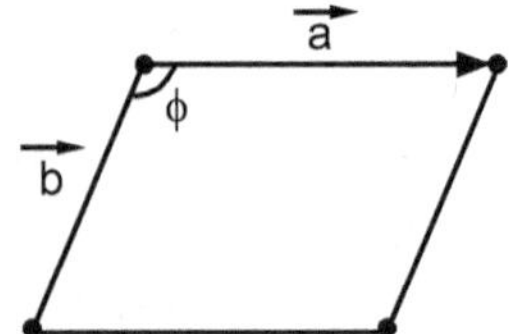

Fig. 1.14 (a) : Oblique (a $\neq$ b, ϕ = 90°)

(b) Rectangular lattice : $|\vec{a}| \neq |\vec{b}|$; $\phi = 90°$

Lattice is invariant under reflection require rectangular lattice.

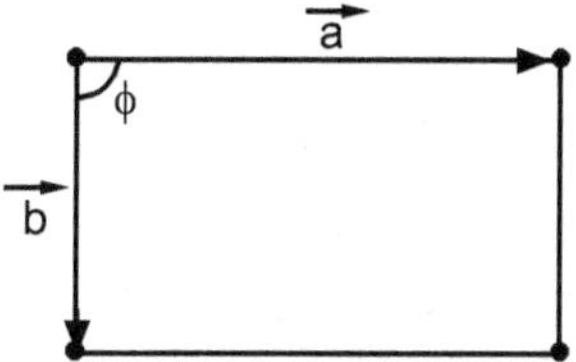

Fig. 1.14 (b) : Rectangular primitive (a ≠ b, ϕ = 90°)

(c) Centered rectangular lattice : $|\vec{a}| \neq |\vec{b}|$; $\phi = 90°$

Invarient under reflection.

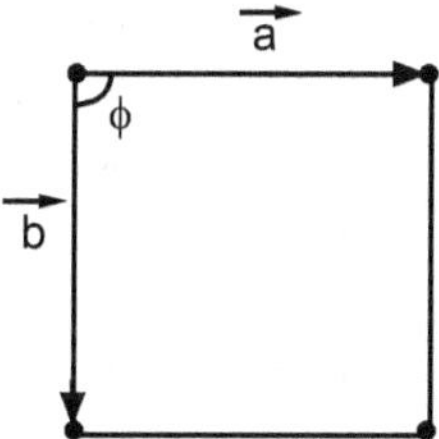

Fig. 1.14 (c) : Rectangular centered (a ≠ b, ϕ = 90°)

(d) Square lattice : $|\vec{a}| = |\vec{b}|$; $\phi = 90°$

The point operation rotation $\dfrac{2\pi}{4}$ require a square lattice.

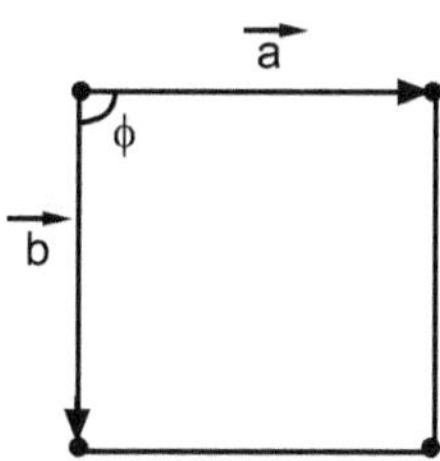

Fig. 1.14 (d) : Square (a = b, ϕ = 90°)

(e) Hexagonal lattice : $|\vec{a}| = |\vec{b}|$; $\phi = 120°$

This is invariant under a rotation $\dfrac{2\pi}{6}$ about an axis through a lattice point and normal to the plane.

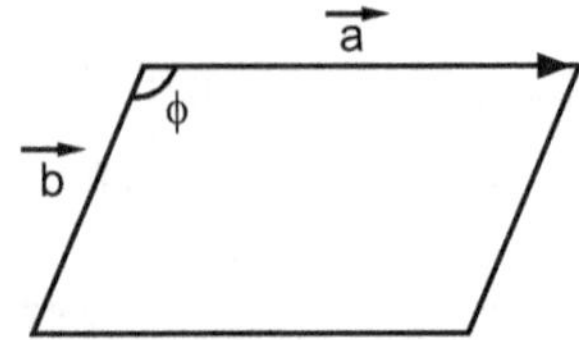

Fig. 1.14 (e) : Hexagonal (a = b, ϕ = 120°)

Bravais lattices in three dimensions :

In three dimensions, point symmetry groups require fourteen different lattice types – one general (triclinic lattice) and thirteen special, collectively called as Bravais lattices or space lattices. There are fourteen ways of arranging points in space lattices such that all the lattice point have exactly the same environment. These fourteen lattice types are conventionally grouped into seven crystal systems, which are characterized by the groups of symmetry elements associated with the lattice points.

Seven crystal system : The seven crystal systems can be distinguished using certain specification about length of the axes and the angle between them. The seven crystal systems are given in terms of their axial relation and angle between them.

1. Cubic　　　　　:　$a = b = c,\ \alpha = \beta = \gamma = 90°$
2. Trigonal　　　　:　$a = b = c,\ \alpha = \beta = \gamma \neq 90° < 120°$
   (Rhombohedral)
3. Hexagonal　　:　$a = b \neq c,\ \alpha = \beta = 90°, \gamma = 120°$
4. Tetragonal　　:　$a = b \neq c,\ \alpha = \beta = \gamma = 90°$
5. Orthorhombic　$a \neq b \neq c,\ \alpha = \beta = \gamma = 90°$
6. Monoclinic　　:　$a \neq b \neq c,\ \alpha = \beta = 90° \neq \gamma$ or $\alpha = \gamma = 90° \neq \beta$
7. Triclinic　　　　:　$a \neq b \neq c,\ \alpha \neq \beta \neq \gamma$

Types of Bravais Lattices (14) :

The lattice symbols are used as

P　-　a primitive lattice　:　It has atoms only at the corners.

I　-　a body centered　:　It has extra atom at the centre.

F　-　a face centered　:　It has atoms at the centre of each face.

C　-　a base centered　:　It has extra atom at the centre of base (or one face centered)

R　-　specific centering position in the trigonal system which gives rhombohedral lattice.

1. Cubic or Regular crystal system : In the cubic system, the crystals are made up of three equal axes at right angles to each other i.e. $a = b = c$ and $\alpha = \beta = \gamma = 90°$. These crystals have three types of lattice depending upon the shape of unit cells. They are :

(a) Simple cubic : In simple cubic, the atoms are only at the corners of the cube. It is primitive cell. e.g. Polonium.

(b) Body centered : In body centered, the atoms are at the corners as well as at the centre of the cube. It is not a primitive cell. e.g. Li, Na, K (alkali metals) and alkaline earth metals.

(c) Face centered : In this case, the atoms are at the corners as well as at the centre of each of the faces of the cube. It is not primitive cell. e.g. NaCl, Ge, Si.

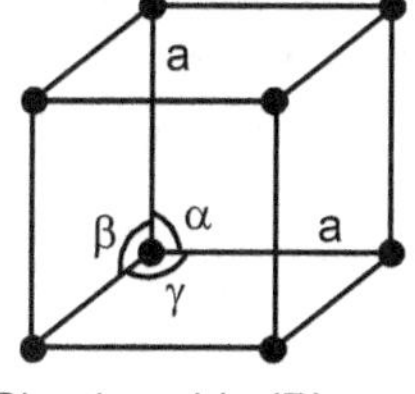
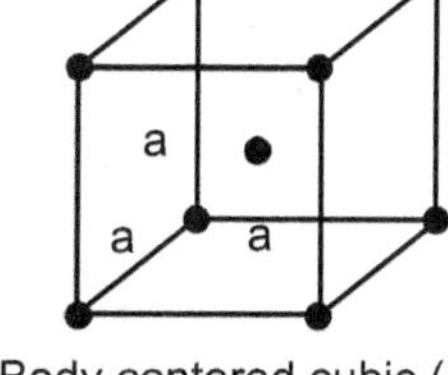
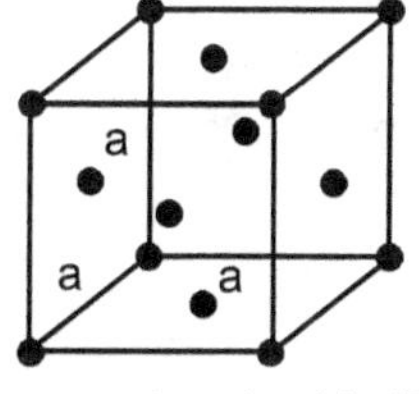

Simple cubic (P)   Body centered cubic (J)   Face-centered cubic (F)

Fig. 1.15

2. Trigonal or Rhombohedral crystal system : In this system, the crystals are made up of three equal axes i.e. $a = b = c$ with equal angles between them but not equal to $90°$ i.e. $\alpha = \beta = \gamma \neq 90°$. Quartz and calcite are examples of this system.

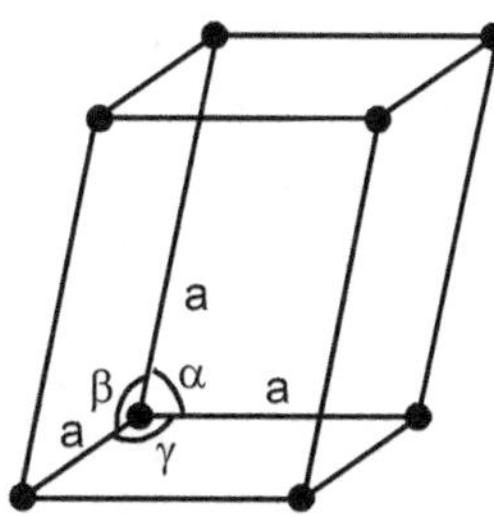

Fig. 1.16 : Simple rhombohedral (R)

3. Hexagonal crystal system : The hexagonal system has eight faces. The two equilateral axes are equal ($a = b$) and intersect at an angle of $\gamma = 120°$. The third vertical axis is of different length ($c \neq a$ or b) at right angles to the equilateral axes, $\alpha = \beta = 90°$. The unit cell is primitive. This type of crystal has 6 fold rotation axes. The examples of this system are Zincite (ZnO), Ice (H_2O), Apatite [$CaCl_2$ $3Ca_2(PO_4)_2$], etc.

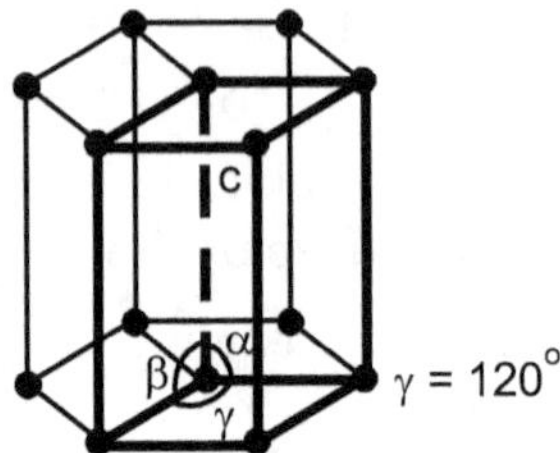

Fig. 1.17 : Simple hexagonal (P)

4.   Tetragonal crystal system : In this system, the three axes are at right angles to each other ($\alpha = \beta = \gamma = 90°$) and the two lateral axes are equal ($a = b \neq c$). The crystal is characterized by a four fold or tetragonal axis of symmetry. It has two types of unit cells, primitive and body centered. e.g. ordinary white tin, indium, cassiterite (SnO_4), Anatase (TiO_2), etc.

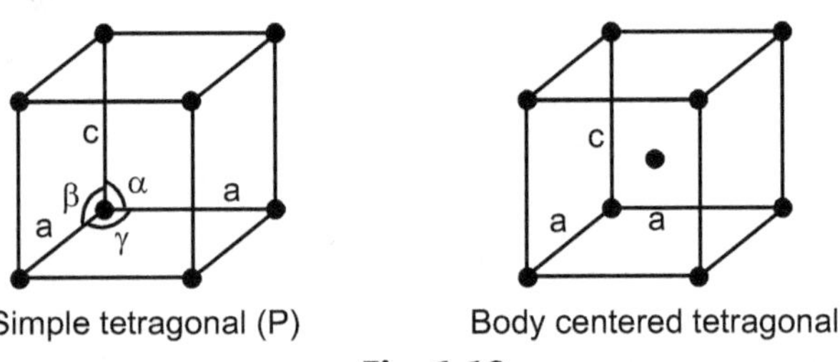

Simple tetragonal (P)   Body centered tetragonal (I)

Fig. 1.18

5.   Orthorhombic crystal system : This system has the crystals in which crystallographic axes are unequal ($a \neq b \neq c$) but they are at right angles to each other ($\alpha = \beta = \gamma = 90°$). There are four types of unit cells (P, C, I and F). There are three mutually perpendicular 2 fold rotation axes. The examples of such crystals are Celestine ($SnSO_4$), Olivine (Mg_2SiO_4), Carnallite (KCl $MgCl_2$, $5H_2O$).

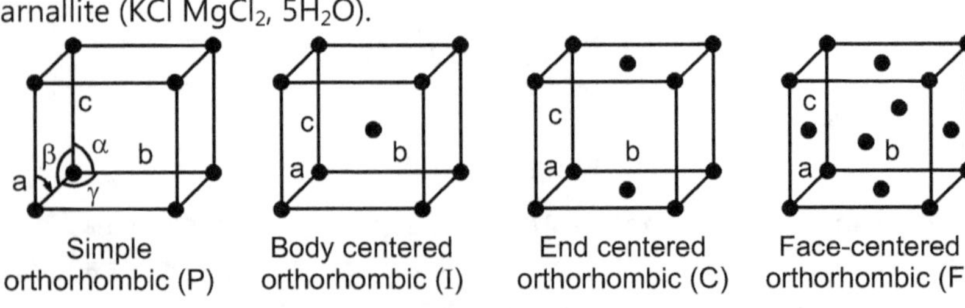

Simple
orthorhombic (P)   Body centered
orthorhombic (I)   End centered
orthorhombic (C)   Face-centered
orthorhombic (F)

Fig. 1.19

6.   Monoclinic or Monosymmetric crystal system : In this type of crystals, the crystallographic axes are unequal (a ≠ b ≠ c) and one axis is perpendicular to the other two axes ($\alpha = \beta = 90°$, $\gamma = 90°$). There are two types of unit cells – one primitive and other base-centered. In this type of crystal there is one two fold rotation axis. Examples of such crystals are Borax, Orthoclase, Cryolite, etc.

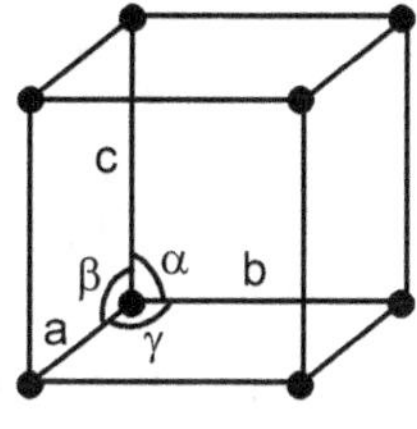

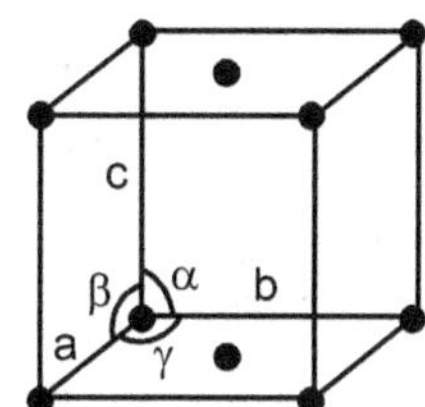

Simple monoclinic (P)   End-centered monoclinic (P)

Fig. 1.20

7.   Triclinic or Asymmetric crystal system : In the triclinic system, neither the three crystallographic axes are equal (a ≠ b ≠ c) nor the three angles are equal ($\alpha \neq \beta \neq \gamma$). Moreover none of the angle being right angle. The unit cell is primitive. Examples are Potassium dichromate ($K_2Cr_2O_7$), Sassolite (H_3BO_3), Albite, Blue vetriol, etc.

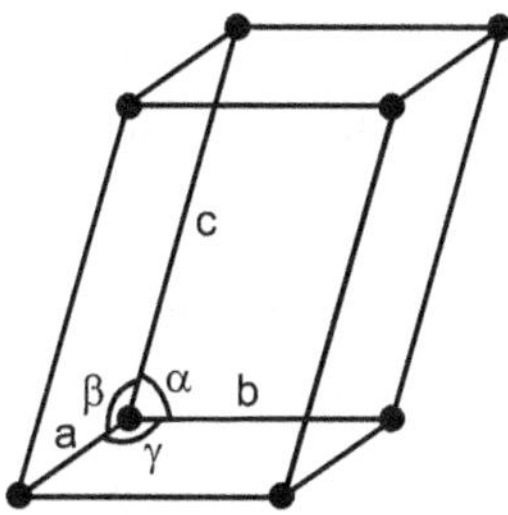

Fig. 1.21 : Simple triclinic (P)

1.8 MILLER INDICES

The lattice points forming a space lattice may be thought of as occupying various sets of parallel planes. The lattice planes can be chosen in a various number of different ways. It is also observed that some planes contain highest density of lattice points (number of lattice points per unit area) and some are weak in lattice point density. In the crystal these planes are defined by giving their orientation without giving their position in space with reference to the direction of the basis vector, plane has a particular orientation, which may be determined by any three points of a plane, provided the points are not collinear.

The scientist Miller arrived at a notation defining the orientation of a plane by three integral numbers h, k, l known as Miller indices. These indices are defined as the reciprocals of the intercepts made by the plane on the three respective axes.

The Miller indices h, k, l for any plane can be obtained by the following procedure.

1. Take any atom in the crystal as origin and erect co-ordinate axes from this atom in the direction of the basis vectors. The axes may be primitive or non-primitive.

2. Find the intercept made by planes on the three crystallographic axes a, b and c.

3. Express the intercepts as multiples of unit distance along the three axes.

4. Take the reciprocals of these intercepts.

5. If fraction result, clear off the fraction by multiplying them by lowest common denominator. This will reduce it to three smallest integers.

Example : Consider set of planes parallel to c axis.

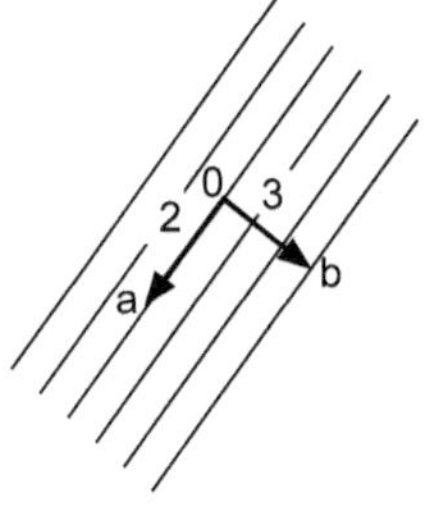

		a	b	c
1.	Axes	a	b	c
2.	Intercepts	2	3	∞
3.	Reciprocal	$\dfrac{1}{2}$	$\dfrac{1}{3}$	$\dfrac{1}{\infty}$
4.	Clear fraction	3	2	0

Fig. 1.22

The resulting three integers (3, 2, 0) are Miller indices of the given plane shown by notations $(\bar{h}, \bar{k}, \bar{l})$ or $[h, k, l]$

In cubic system this scheme has several features as :

For cubic unit cell if origin is at the corner and the axes are parallel to edges,

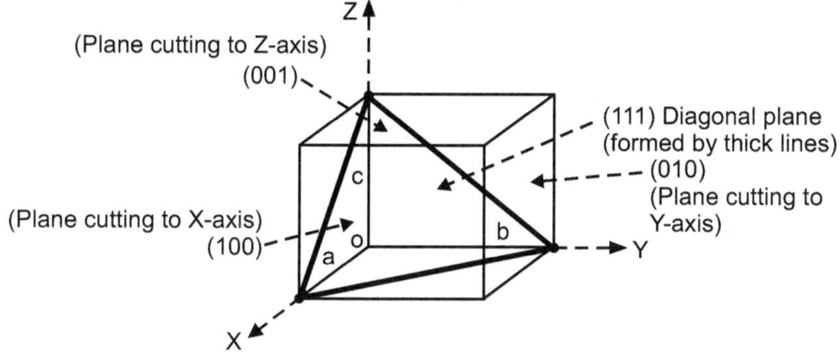

Fig. 1.23 : Cubic system

1. The planes which are parallel to each other have the same orientation and hence have same Miller indices.

2. If the plane is parallel to the crystal axis, the intercept made by this plane with that axis will be infinite and corresponding Miller index will be equal to zero, since reciprocal of infinite is zero.

3. A plane passing through which origin is defined as in terms of a parallel plane having non-zero intercepts.

4. The Miller indices will be negative if plane cuts the negative sides

   of axes e.g. $(\bar{h}, \bar{k}, \bar{l})$ or $(\bar{h}, k, l)$, $(h, \bar{k}, l)$ or $(h, k, \bar{l})$.

5. The angle θ between two crystallographic directions $(h\ k\ l)$ and $(h'\ k'\ l')$ is given by

$$\cos \theta = \frac{hh' + kk' + ll'}{(h^2 + k^2 + l^2)^{1/2} (h'^2 + k'^2 + l'^2)^{1/2}}$$

6. The normal to the plane with index number $(h\ k\ l)$ is the direction $[h\ k\ l]$. That is the normal to a plane is a direction with the same indices as the plane.

1.9 INTERPLANER SPACING AND MILLER INDICES

Consider a family of parallel planes (h, k, l) in a cell in which coordinate axes are orthogonal. The spacing d between the adjacent planes of this family can be calculated by taking any lattice point as an origin, through which reference plane passes. Then erect coordinate axes in the a, b and c directions (i.e. x, y and z directions) and find the perpendicular distance between this origin and the point P in the plane ABC (Refer Fig. 1.20), which is nearest to this origin. This plane would obviously have $\frac{a}{h},\frac{b}{k}$ and $\frac{c}{l}$ as intercepts on the a, b and c axes (i.e. X, Y and Z axes) respectively. Here 'd' is the length of the normal to the plane ABC from origin O and 'd' will be the distance between adjacent planes i.e. interplaner spacing.

We have to find an expression for 'd' in terms of a, b, c and h, k, l.

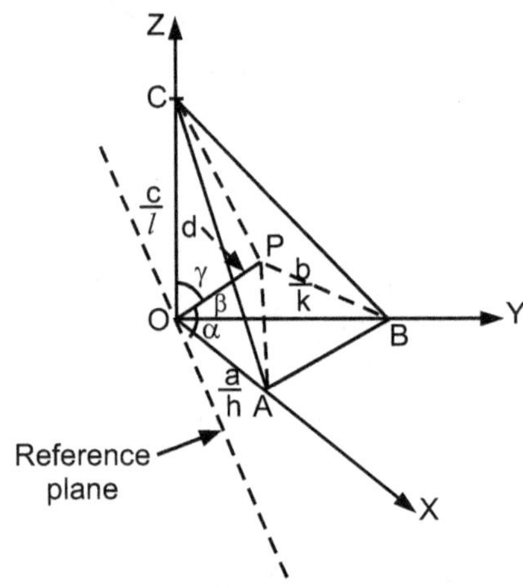

Fig. 1.24 : The distance between adjacent planes

Now referring to Fig. 1.24, it is evident that

$$\angle\, OPA \;=\; \angle\, OPB = \angle\, OPC = 90^\circ$$

From right angled triangle OPA,

$$\cos \alpha \;=\; \frac{d}{a/h} = \frac{d\cdot h}{a}$$

the direction of cosine of the normal along OX.

Similarly from triangle OPB and triangle OPC, we have,

$$\cos \beta \;=\; \frac{d}{b/k} = \frac{dk}{b} \quad \text{and} \quad \cos \gamma = \frac{d}{c/l} = \frac{dl}{c}$$

where cos β and cos γ are the direction cosines of the normal along OY and OZ direction respectively.

Then according to cosine rule, we can write

$$\cos^2 \alpha + \cos^2 \beta + \cos^2 \gamma = 1$$

$$\therefore \quad \frac{d^2 h^2}{a^2} + \frac{d^2 k^2}{b^2} + \frac{d^2 l^2}{c^2} = 1$$

$$\therefore \quad d^2 \left(\frac{h^2}{a^2} + \frac{k^2}{b^2} + \frac{l^2}{c^2} \right) = 1$$

$$\therefore \quad d = \frac{1}{\left(\dfrac{h^2}{a^2} + \dfrac{k^2}{b^2} + \dfrac{l^2}{c^2} \right)^{1/2}} = \frac{1}{\sqrt{\dfrac{h^2}{a^2} + \dfrac{k^2}{b^2} + \dfrac{l^2}{c^2}}}$$

Thus in general the interplaner distance (spacing) for the plane having Miller indices (h k l) in orthogonal (axes are mutually perpendicular) systems for primitive lattice is given by

$$d_{hkl} = \frac{1}{\sqrt{\dfrac{h^2}{a^2} + \dfrac{k^2}{b^2} + \dfrac{l^2}{c^2}}}$$

From this equation, it is seen that the distance between parallel planes decreases as the indices increases. Thus high index reflection require X-rays of shorter wavelength.

Cases :

1. For cubic system, a = b = c.

$$\therefore \quad d_{hkl} = \frac{a}{\sqrt{h^2 + k^2 + l^2}}$$

2. For tetragonal system, a = b ≠ c.

$$\therefore \quad d_{hkl} = \frac{1}{\sqrt{\left(\dfrac{h^2 + k^2}{a^2} \right) + \dfrac{l^2}{c^2}}}$$

3. For orthorhombic system, a ≠ b ≠ c.

$$\therefore \quad d_{hkl} = \frac{1}{\sqrt{\dfrac{h^2}{a^2} + \dfrac{k^2}{b^2} + \dfrac{l^2}{c^2}}}$$

1.10 SIMPLE CRYSTAL STRUCTURES
Cubic (SC, BCC, FCC)
and Hexagonal Close Packed (HCP)

The crystal structures of sc, bcc, fcc and hcp are explained with respect to coordination number, atomic radius, atoms per unit cell and packing fraction.

1. Coordination number : The coordination number is defined as the number of equidistance nearest neighbours that an atom has in the given crystal structure. Greater the coordination number, the more closely packed will be the crystal structure.

2. Atomic radius (r) : In crystal structures the atoms are supposed to be perfect spheres in contact and hence the distance between the centres of atoms is equal to the diameter of the atom. Atomic radius is defined as half the distance between the neighbours in crystal of pure element. Usually atomic radius is expressed in terms of the cube edge (a).

3. Packing fraction or Packing density : The packing fraction (or factor) is defined as the ratio of volume of atoms occupying the unit cell to the volume of the unit cell relating to that structure.

$$\therefore \quad \text{Packing fraction} = \frac{\text{Volume of atoms per unit cell (v)}}{\text{Volume of the unit cell (V)}}$$

It is always less than one. Larger the packing fraction, more closed will be the crystal structure.

(i) Simple cubic structure (SC) :

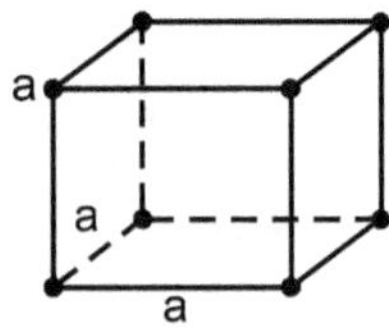

Fig. 1.25 : Unit cell of simple cubic structure

In this structure the atoms are situated at the corners of the cube touching each other along the edges.

Each atom in such structure is surrounded by six neighbours at the same distance each equal to the length of unit cell, so that the coordination number is 4 + 2 = 6 i.e. 4 - nearest in same plane and 2 - in a vertical plane (one above and one below).

In simple cubic structure, eight atoms are placed at the eight corners of a cubic unit cell. Since each atom is common to the corners of 8 cubes, the contribution of each atom to the unit cell is $\frac{1}{8}$ and as there are 8 corners in each unit cell, the number of atoms per unit cell is $\frac{1}{8} \times 8 = 1$.

This structure is loosely packed and rarely found in nature.

e.g. Polonium (Po) at certain temperature.

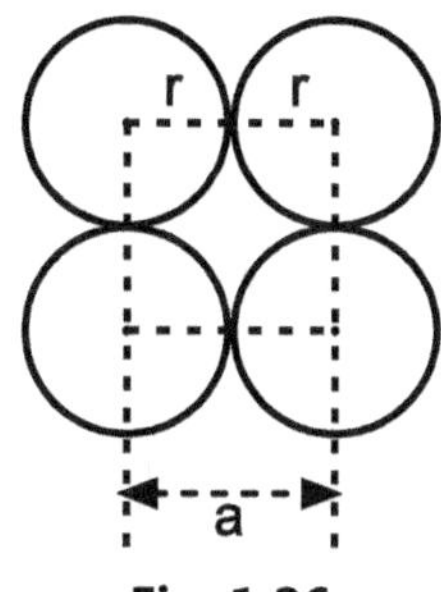

Fig. 1.26

If 'a' is the side of the unit cell, then distance between nearest neighbours is 'a'.

Since $2r = a$, atomic radius $r = \frac{a}{2}$.

$$\text{Volume of atom} = v = \frac{4}{3}\pi r^3 = \frac{4}{3}\pi\left(\frac{a}{2}\right)^3 = \frac{4a^3}{6}$$

and Volume of unit cell $= V = a \times a \times a = a^3$

$$\therefore \quad \text{Packing fraction} = \frac{v}{V} = \frac{\pi a^3}{6a^3} = \frac{\pi}{6} = \frac{3.142}{6} = 0.5236$$

(ii) Body centered cubic structure (BCC) : A bcc lattice can be considered to contain two identical simple lattices, one consist of the eight corner atoms and the other consist of only centre atom in cube. In other words, unit cell of BCC is a cubic, in which one atom is at each corner and one atom in the centre of the cube.

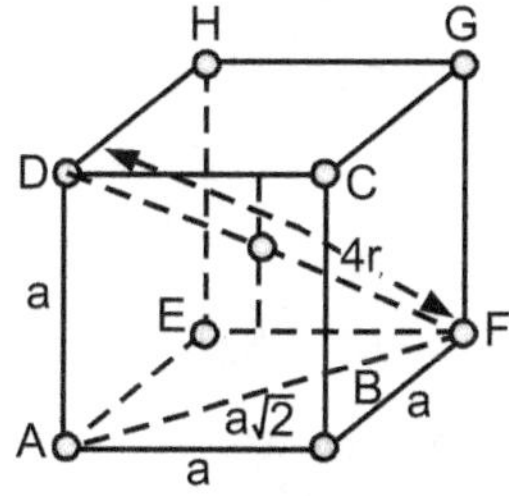

Fig. 1.27 : Unit cell of the BCC structure

Now, since a corner atom is shared among eight cells which touch there and central atom is totally shared by unit cell, there are $\left(\dfrac{1}{8}\right) \times 8 + 1 = 2$ atoms associated with this cell. The cell is therefore a non-primitive one.

This structure is loosely packed, since each atom has only eight nearest neighbours i.e. coordination number is 8, but the atoms are in contact along body diagonals. The number of atoms per unit length is more along diagonal than cubic edge.

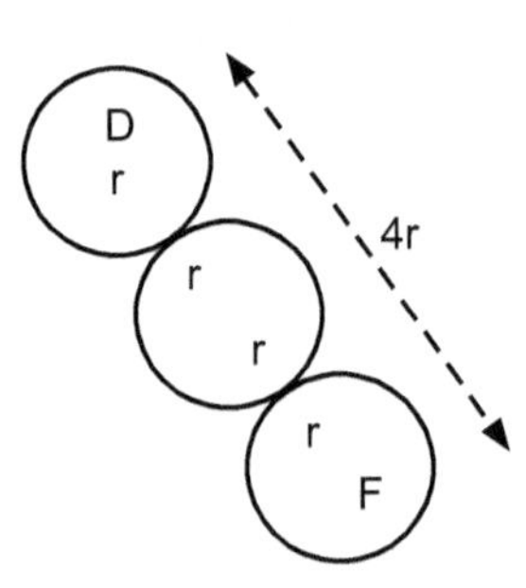

Fig. 1.28

From Fig. 1.28, in Δ DAF,

$$(DF)^2 = (DA)^2 + (AF)^2$$

$$\therefore (4r)^2 = a^2 + (a\sqrt{2})^2$$

From Δ ABF, we get AF = $a\sqrt{2}$

$$\therefore (4r)^2 = 3a^2$$

$$\therefore r^2 = \frac{3a^2}{16}$$

$$\text{or} \quad r = \frac{\sqrt{3}}{4}a$$

The atomic radius 'r' is given by half the distance between nearest neighbours in a crystal.

$$\therefore \quad \text{nearest distance} = 2r = \frac{\sqrt{3}}{2}a$$

The volume of 2 atoms in the unit cell,

$$v = 2 \times \frac{4}{3}\pi r^3 = \frac{4a^3\sqrt{3}}{8}$$

and the volume of the unit cell,

$$V = a \times a \times a = a^3$$

$$\therefore \quad \text{Packing fraction} = \frac{v}{V} = \frac{\pi a^3 \sqrt{3}}{8a^3} = \frac{\pi\sqrt{3}}{8} = \frac{3.14 \times \sqrt{3}}{8} = 0.68$$

Elements that possess body centered cubic structure are Na, K, (alkali metals) and Cr, Mo, W, etc. In BCC structure, repetition sequence is ABAC ABAC ......

(iii) Face centered cubic structure (FCC) : The unit cell FCC structure is shown in Fig. 1.29. It is cubic and an atom occupies each corner of the cube and the centre of each face. FCC has 4-fold rotation axis of symmetry about any one of cube edges. Thus each atom having 12 nearest neighbours. These structures are most closely packed, since coordination number is 4 + 4 + 4 = 12.

i.e. 4 FCC atoms surrounding unit cells in its own plane,

4 FCC atoms below this plane,

4 FCC atoms above this plane.

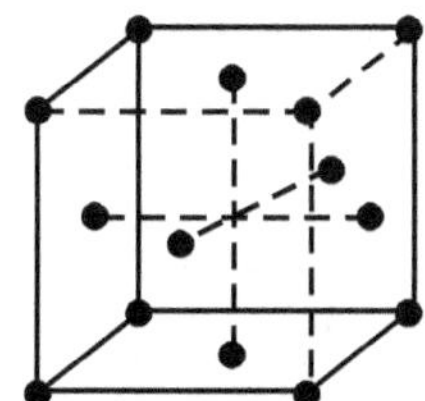

Fig. 1.29 : Unit cell of FCC structure

In this case, the atoms at the corners are shared by eight surrounding unit cells and each of face centered atoms are shared by 2 surrounding unit cells.

Therefore, total number of atoms in FCC unit cell

$$= \frac{1}{8} \times 8 + \frac{1}{2} \times 6 = 1 + 3 = 4$$

In FCC structure, the atoms are in contact along the diagonal of the faces.

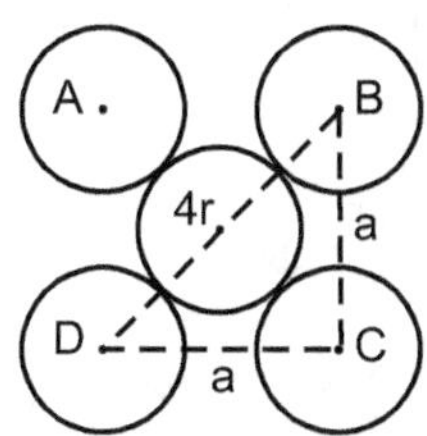

Fig. 1.30

From Fig. 1.30,

$$(DB)^2 = (DC)^2 + (CB)^2$$
$$= a^2 + a^2$$
$$(4r)^2 = 2a^2$$

$\therefore$ Atomic radius, $r = \dfrac{\sqrt{2}}{4} a$

and nearest distance, $2r = \dfrac{\sqrt{2}}{2} a = \dfrac{a}{\sqrt{2}}$

The volume of 4 atoms in the unit cell,

$$v = 4 \times \frac{4}{3} \pi r^3 = 4 \times \frac{4}{3} \pi \left(\frac{\sqrt{2}}{4} a \right)^3$$

$$= \frac{\pi a^3 \sqrt{2}}{6}$$

The volume of the unit cell,

$$V = a \times a \times a = V^3$$

$\therefore$ Packing fraction $= \dfrac{v}{V} = \dfrac{\pi a^3 \sqrt{2}}{6a^3} = \dfrac{\pi}{3\sqrt{2}} = 0.74$

Elements possessing this structure are Cu, Ag, Au, Ca, Al, Pb and Pt. In FCC, repetition sequence is ABC ABC ABC.

(iv) Hexagonal close packed structure (HCP) : The hexagonal close packed structure is made by stacking close packed planes in a simple sequence.

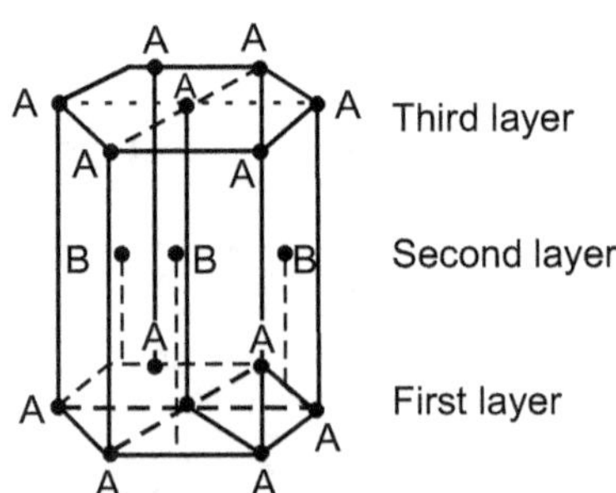

Fig. 1.31 : Hexagonal close packed structure

In this structure, every atom is supposed to be in the form of a sphere and are arranged in such a way that they occupy minimum space. There are two ways of arranging identical spheres, so that they occupy minimum space. One way is a structure with cubic symmetry i.e. face centered cubic and the other is hexagonal symmetry i.e. hexagonal close packed structure.

In the first layer, sphere may be arranged in a single close packed layer by placing each sphere in contact with six others.

A second similar layer may be placed on the top of this by placing each sphere in contact with three spheres of bottom layer i.e. in second layer spheres can be arranged in alternate depression or valleys in the first layer marked as B.

A third layer can be added in two ways of arranging the sphere (a) the sphere can be arranged vertically above the sphere in the first layer, thus the sequence of arranging the sphere or the stacking becomes AB AB AB ...... This kind of packing of spheres is called the hexagonal close packing (Refer Fig. 1.31 and Fig. 1.32).

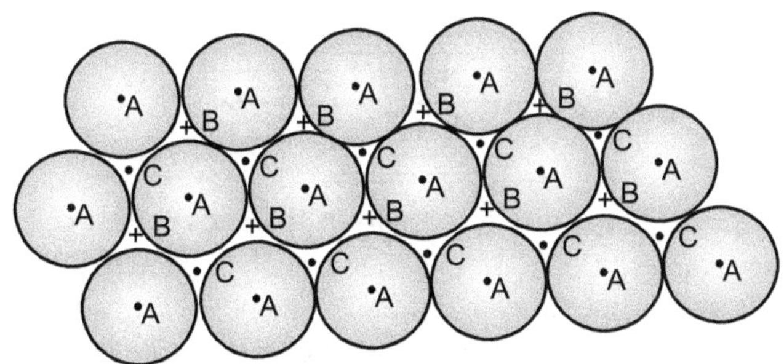

Fig. 1.32 : Close packing of spheres

(b) The other way of arranging spheres in third layer is by placing the spheres over the position 'C' i.e. spheres in the third layer are placed over the depression or valleys in the first layer which are not occupied by the spheres in the second layer. This sequence of arranging or stacking of spheres is ABC ABC ...... This kind of packing of spheres is called cubic close packing (FCC).

Both sequence of packing represent the dense packing of spheres.

The atom position in this structure do not constitute a space lattice since they do not transform into one another by the simple symmetry operation of replacing a vector r from an origin by − r. The space lattice is simple hexagonal with two identical atoms associated with each lattice point.

The hexagonal unit cell can be divided into three rhombohedral unit cells as shown in Fig. 1.33. Each rhombohedral cell has atoms at the eight corners and one atom between the bottom and top layer that is at a height of $\frac{c}{2}$ from the bottom plane, where c is the lattice parameter on the c-axis. Thus it contains two atoms per unit cell. The conventional cell contain six atoms.

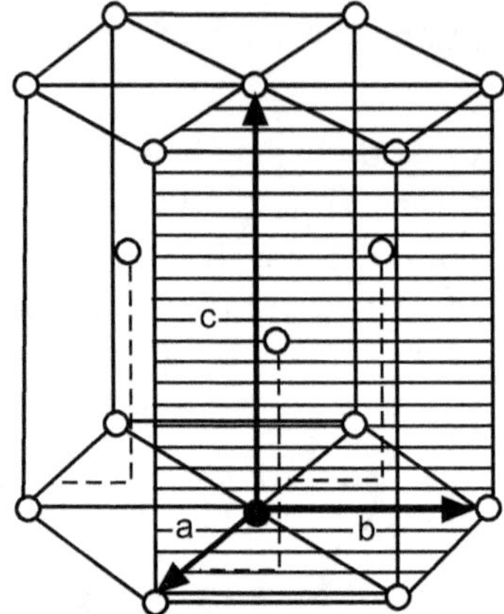

Fig. 1.33 : Primitive cell within the convention cell

Primitive cell, which is simple hexagonal is shown by shading. It contains two atoms per lattice point (dark circle).

In hexagonal close packed structure, each atom has 12 nearest neighbours, 6 in its plane, 3 in the above and 3 in the plane below. Hence coordination number is 12.

Determination of $\frac{c}{a}$ ratio :

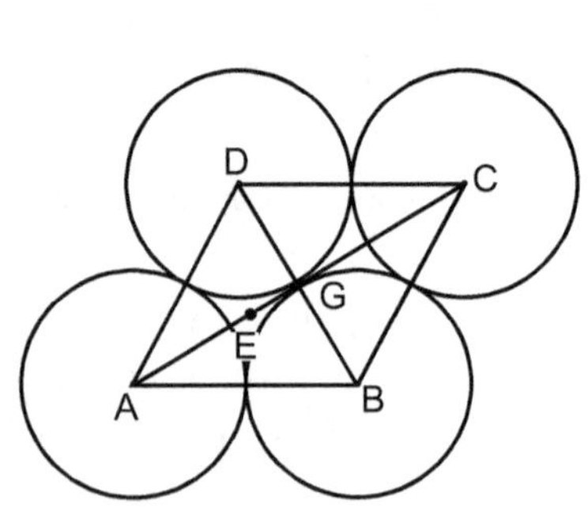

(a) Base of rhombohedral unit cell

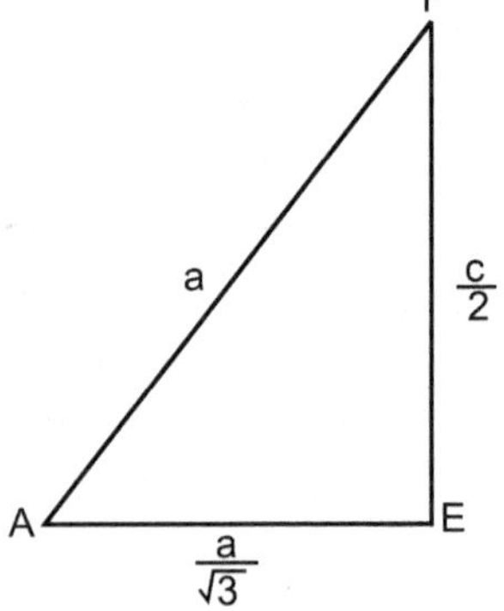

(b) Position of B atom

Fig. 1.34

In Fig. 1.34 (a), the base of a rhombohedral unit cell shown, in which

$AD = AB = CD = BC = a$ and $DG = \dfrac{a}{2}$.

The sphere (not shown) in the next layer has its centre at F vertically above E and it touches the three spheres whose centres are A, B and D. [Refer Fig. 1.34 (b)].

In Fig. 1.34 (a), ABD is an equilateral triangle and AG is one of its median. Then

$$AE = \frac{2}{3} \text{ of median } AG$$

$$AE = \frac{2}{3}\left(\sqrt{a^2 - \frac{a^2}{4}}\right) \qquad \text{(since in } \Delta \text{ AGB)}$$

$$\therefore \qquad AE = \frac{2}{3} \cdot \frac{a\sqrt{3}}{2} = \frac{a}{\sqrt{3}}$$

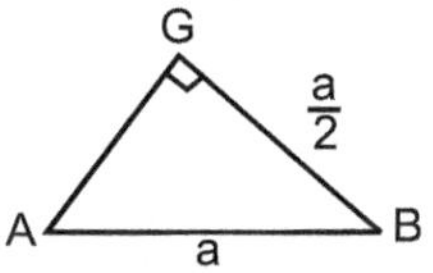

Fig. 1.35

From Δ AEF, $(AF)^2 = (AE)^2 + (FE)^2$

$\therefore$ $(FE)^2 = (AF)^2 - (AE)^2$

$$= a^2 - \left(\frac{a}{\sqrt{3}}\right)^2$$

$$\therefore \qquad FE = \sqrt{a^2 - \frac{a^2}{3}} = a\sqrt{\frac{2}{3}}$$

Also $FE = \dfrac{c}{2}$

$$\therefore \qquad \frac{c}{2} = a\sqrt{\frac{2}{3}} \quad \text{or} \quad \frac{c}{a} = 2\sqrt{\frac{2}{3}} = \sqrt{\frac{8}{3}}$$

$$\therefore \qquad \frac{c}{a} = 1.633$$

In case of close packing structure, $a = 2r$, where r is the radius of sphere representing an atom.

$$\text{Atomic radius} = r = \frac{a}{2}$$

Volume of 2 atoms in the cell,

$$v = 2 \times \frac{4}{3} \pi r^3$$

$$\therefore \quad v = \frac{8}{3} \pi \left(\frac{a}{2}\right)^3 = \frac{\pi a^3}{3}$$

Volume of rhombohedral unit cell,

$$V = \frac{\sqrt{3}}{2} a^2 c \qquad \left(\text{since } \frac{a\sqrt{3}}{2} \text{ - median AG}\right)$$

$$\therefore \quad \text{Packing fraction} = \frac{v}{V} = \frac{\pi a^3/3}{\frac{\sqrt{3}}{2} a^2 c} = \frac{2\pi a}{3\sqrt{3}\,c} = \frac{2\pi}{3\sqrt{3}} \left(\frac{a}{c}\right)$$

$$= \frac{2 \times 3.14}{3\sqrt{3}} \left(\frac{1}{1.633}\right) \qquad \cdots \frac{a}{c} = \frac{1}{1.633}$$

$$\text{Packing fraction} = 0.74$$

The HCP crystal structure possesses maximum packing density. Examples of HCP structure are Be, Mg, Zn, He, Ti, Co, Cd, Y, Zr, Gd, Lu, Ru, etc.

SOLVED PROBLEMS

Problem 1.1 : Lead is face centered cubic with an atomic radius of r = 1.746 A°. Find the spacing of (200) plane and (220) plane.

Solution : In FCC crystal structure, the atomic radius r of a unit cell is related to the lattice constant a by the relation

$$a = \frac{4r}{\sqrt{2}} A° = \frac{4 \times 1.746}{\sqrt{2}} = 4.93 \ A°$$

The interplaner spacing is given by

$$d_{hkl} = \frac{a}{\sqrt{h^2 + k^2 + l^2}}$$

(i)   For (200) plane, h = 2, k = 0, l = 0

$$\therefore \quad d_{200} = \frac{4.93}{\sqrt{(2^2 + 0^2 + 0^2)}} = \mathbf{2.465 \ A°}$$

(ii)   For (220) plane, h = 2, k = 2, l = 0

$$\therefore \quad d_{220} = \frac{4.93}{\sqrt{(2^2 + 2^2 + 0^2)}} = \mathbf{1.74 \ A°}$$

Problem 1.2 : The lattice constant of a cubic lattice is a = 2.5 A°. Calculate the spacing between (100), (110) and (111) planes.

Solution : In cubic lattice, spacing d_{hkl} between planes is given by

$$d_{hkl} = \frac{a}{\sqrt{h^2 + k^2 + l^2}}$$

(i) For plane (100), h = 1, k = 0, l = 0

$$\therefore \quad d_{100} = \frac{2.5}{\sqrt{(1^2 + 0^2 + 0^2)}} = \textbf{2.5 A}^\circ$$

(ii) For (110) plane, h = 1, k = 1, l = 0

$$\therefore \quad d_{110} = \frac{2.5}{\sqrt{(1^2 + 1^2 + 0^2)}} = \frac{\textbf{2.5}}{\sqrt{\textbf{2}}} \textbf{ A}^\circ$$

(iii) For (111) plane, h = 1, k = 1, l = 0

$$\therefore \quad d_{111} = \frac{2.5}{\sqrt{(1^2 + 1^2 + 1^2)}} = \frac{\textbf{2.5}}{\sqrt{\textbf{3}}} \textbf{ A}^\circ$$

EXERCISES

(A) Multiple Choice Type Questions :

1. The number of point group and lattice structure in three dimensions are ......

    (a) 7, 14 **(b) 32, 14**

    (c) 14, 7 (d) 14, 32

2. The packing fraction for FCC crystal structure is ......

    (a) 0.52 **(b) 0.74**

    (c) 0.68 (d) 0.25

3. The atomic radius of simple cubic lattice is ......

    (a) a **(b) $\dfrac{a}{2}$**

    (c) $\sqrt{2}\, a$ (d) 2a

4. The Miller index for the plane which does not cut the crystallographic axis is ......

    (a) 0 (b) 1

    (c) $\bar{1}$ (d) ∞

5. The closely packed structures are ......

(a) SC and FCC (b) BCC and FCC

(c) HCP and FCC (d) BCC and HCP

6. The interplaner distance for (221) plane in case of cubic lattice is ......

(a) $\dfrac{a}{5}$ (b) $\dfrac{a}{9}$

(c) $\dfrac{a}{3}$ (d) $\dfrac{a}{6}$

7. Primitive unit cell contains ...... number of atoms.

(a) one (b) two

(c) three (d) four

8. The Miller index for the plane which cut the X-axis at $\dfrac{1}{3}\bar{a}$ is ......

(a) 3 (b) 1

(c) $\bar{3}$ (d) $\bar{1}$

9. The number of lattice points per unit cell for non-primitive unit cell is ......

(a) equal to 1 **(b) greater than 1**

(c) less than 1 (d) none of these

10. The coordination number for FCC lattice is ......

(a) 6 (b) 8

(c) 12 (d) 14

11. The atomic radius of FCC crystal is ......

(a) $\dfrac{a\sqrt{2}}{4}$ (b) $a\dfrac{\sqrt{3}}{4}$

(c) $a\dfrac{\sqrt{5}}{4}$ (d) $a\dfrac{4}{\sqrt{2}}$

12. Miller index of the plane parallel to crystallographic axes is ......

(a) 0 (b) 1

(c) ∞ (d) none of these

13. A crystal system with $a = b \neq c$ and $\alpha = \beta = \gamma = 90°$ is called ......

(a) trigonal **(b) tetragonal**

(c) monoclinic (d) cubic

14. The sequence of atoms in hexagonal close packed crystal structure is ......

(a) ABC ABC **(b) AB AB**

(c) random (d) ABCD ABCD

15. If $\dfrac{360°}{n}$ gives n-fold axis of symmetry then for diad axis, the cube must be rotated through the angle of ......

(a) $60°$ (b) $90°$

(c) $120°$ **(d) $180°$**

16. Packing fraction of HCP crystal structure is ......

(a) 0.68 (b) 0.32

(c) 0.74 (d) 0.52

17. The Miller indices of a plane which cut the X-axis at half way of the unit cell are given by ......

(a) (020) (b) (002)

(c) (200) (d) $\left(\dfrac{1}{2}\,00\right)$

(B) Short Answer Type Questions :

1.   Write a short note on crystal structure.

2.   Define packing fraction and find its value for BCC crystal.

3.   Write a short note on Bravais lattices.

4.   Explain the terms coordination number and packing fraction for crystal structure. Find the expression for packing fraction of a simple cubic crystal structure.

5.   Explain how Miller indices for an atomic plane are obtained. Show (111) plane for cubic crystal structure.

6.   Define packing fraction. Find relation for the same in case of FCC crystal.

7.   Define and explain the terms for FCC crystal :

(a) Coordination number

(b) Packing fraction

(c) Atomic radius

8. Explain nomenclature of atomic planes by Miller indices.

9. Distinguish between crystalline and amorphous substances.

10. Prove that the packing density of hcp and FCC crystal structure is the same.

11. Explain two dimensional Bravais lattice types.

12. Explain three dimensional Bravais lattice types.

13. Explain the terms lattice, basis and crystal structure.

(C) Long Answer Type Questions :

1. Discuss symmetry elements for a cube.

2. Derive an expression for the interplaner distance between the lattice planes of cubic system.

3. Define space lattice and describe fourteen types of Bravais lattices.

4. What are Miller indices ? How are they obtained ? Show (200) plane for simple cubic crystal.

5. Show that packing fraction of FCC and HCP crystal structure is the same.

6. Obtain a expression for interplaner spacing of (hk$l$) planes for cubic crystal.

(D) Problems for Practice :

1. Calculate the lattice spacing for (111) planes in a orthorhombic space group, where a = 2.4 A°, b = 3.1 A° and c = 1.9 A°.

2. For a FCC crystal atomic radius is 1.72 A°. Find the interplaner spacing for (211) and (220) planes.

3. For an FCC crystal with atomic radius 1.653 A°, find the spacing between (223) plane.

4. Determine Miller indices for planes (3a, 6b, 2c) and (2a, 4b, $\frac{c}{2}$).

5. The lattice constant of a cubic lattice is a = 2.814 A°. Calculate the spacing between (011) and (101) planes.

2

CHAPTER

X-RAY DIFFRACTION BY CRYSTALS

SYLLABUS

Reciprocal lattice, Properties of reciprocal lattice, Bragg's law in reciprocal lattice (Ewald's construction), Powder method of X-ray diffraction and analysis of cubic crystal structure.

2.1 INTRODUCTION

In the past, people studied crystals through their geometry, that is, measurement of angles between crystal faces and thus, they established the symmetry of crystals. This kind of study was carried out with the help of a goniometer. However, the internal structure of crystals could not be revealed. Then, people started using optical microscopes that can also could not reveal the internal structure of the crystal but this technique proved to be very useful for studying different aspects of biological samples. Then, people tried diffraction of electromagnetic waves through the specimen and method of getting information about the structure of the specimen was through the analysis of diffraction patterns that emerged from the sample. Ordinary light can be diffracted through grating producing diffraction pattern. However, ordinary light passing through a crystal cannot show the diffraction pattern.

In X-ray diffraction, the beams of X-rays fall on different atoms which are arranged by a set of parallel planes called as lattice planes in the crystal. As we know that in, crystallography, a family of parallel planes with the same interplaner spacing between them are considered and are represented by the same Miller indices. But a considerable simplification can be done if the crystal planes are described by their normals and interfacial spacing as it is much easier to visualize one dimensional normals than two dimensional planes. In old crystallographic methods, we can determine only about the shape of unit cell of the crystal. But with X-ray diffraction technique it becomes possible to prove shape as well as size of the unit cell of the crystal, so the crystallography is concerned itself not only with the orientation of the planes, but also with their spacings. A useful projection should present both the orientations and the spacing of the planes. A new projection was developed which displays the orientations as well as spacing of the planes. With this idea, reciprocal lattice is developed. The Bragg's condition of X-ray diffraction can be well explained by the concept of reciprocal lattice.

In this chapter, we study the concept of reciprocal lattice along with their properties. Also we study powder method by X-ray diffraction technique and analysis of cubic crystal structure.

2.2 RECIPROCAL LATTICE

The idea of reciprocal lattice was proposed by P. P. Ewald in 1921.

The concept of reciprocal lattice is important because it is easier to picture a collection of points in reciprocal space than it is to visualize a complex of interpenetrating planes in crystal. In the crystal lattice there exist many set of planes with different orientations and spacing which can cause diffraction. A two dimensional plane can be very well represented by its normal which has only one dimension. The direction of normal represents the orientation of plane. If the length assigned to each normal is proportional to the reciprocal of the interplaner spacing of that plane. The points at the end of the normals drawn from a common origin form a lattice which is called as reciprocal lattice. There are two types of lattices, the first one is the direct lattice or Bravais lattice defined by positions of the lattice points which are related to the positions of atoms called as unit cells and the other is the reciprocal lattice related to the unit cell of the reciprocal lattice known as Brillouin zone.

The collection of such points represents (i) a collection of slopes of the direct lattice planes indicated by the slopes of the normals and (ii) a collection of interplanar spacings of the direct lattice in terms of reciprocal spacings. The collection of the points is a lattice array and thus constitutes a new lattice, the reciprocal lattice.

2.3 GEOMETRICAL CONSTRUCTION OF RECIPROCAL LATTICE-ONE DIMENSIONAL LATTICE

Consider a family of parallel and equidistant lattice planes perpendicular to axis say (100) as shown in Fig. 2.1, they will be infinite in number since (200), (300), (400) ... planes are equivalent planes.

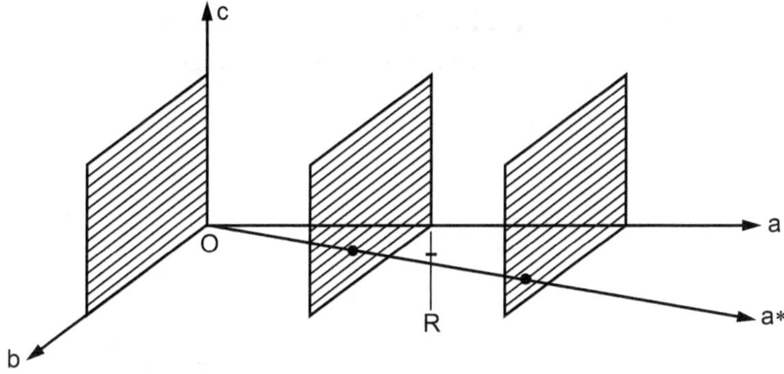

Fig. 2.1

The draw the reciprocal points for these planes, draw a common normal from origin 'O' taken arbitrarily on any lattice point and locating a point on this at a distance which is inversely proportional to the interplanar spacing. In Fig. 2.1, OP represents common normal and R_1 represents reciprocal lattice, such that $OR \propto \dfrac{1}{d_{100}}$ i.e. $OR_1 = \dfrac{M}{d_{100}}$, where M is constant of proportionality called as scale factor for OR_1, which defines the reciprocal point.

Similarly we find reciprocal points for (200), (300), (400) planes parallel to (100) plane and having spacing equal to $\dfrac{1}{2}, \dfrac{1}{3}, \dfrac{1}{4}$ ... the spacing of the (100) plane. According to definition, the reciprocal point for (200) plane along OP is R_2, such that,

$$OR_2 = \frac{M}{d_{100}/2} = \frac{2M}{d_{100}} = R_2$$

The reciprocal lattice for (300) plane is

$$OR_3 = \frac{M}{d_{100}/3} = \frac{3M}{d_{100}} = R_3$$

and so on. Thus we find that all the planes are represented by series of reciprocal points R_1, R_2, R_3 ... and so on which are situated along a single line and these reciprocal points are equally spaced. These points form a one-dimensional lattice.

2.4 GRAPHICAL OR GEOMETRICAL DEMONSTRATION OF RECIPROCAL LATTICE IN TWO DIMENSIONS

Let us take a monoclinic crystal and construct it as reciprocal lattice.

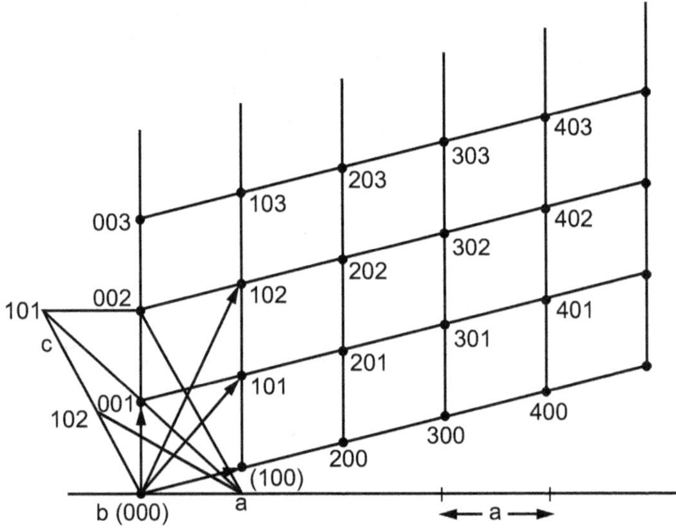

Fig. 2.2

Let us consider (100) plane in the monoclinic crystal. This plane will be parallel to b-axis. Normal to this plane will lie in the plane of paper, simply by considering that (100) plane is parallel to the b-axis or vertical axis and so normal to it must lie in the plane of paper.

To construct the reciprocal lattice to the direct monoclinic lattice, first origin is selected which is at the origin of the a-b and c-axis. Normal is drawn from the origin to (100) plane in the direction as shown in Fig. 2.2. Now lines are drawn parallel to the normal to (100) planes from 001, 002, 003 ... points. In addition to this the lines are drawn parallel to the normal to (001) plane from 100, 200, 300 ... points, so the resulting figure will be a collection of intersection points between the two sets of parallel lines. Refer Fig. 2.2. The collection of intersection points is the reciprocal lattice.

Each intersection point represents a plane. For example, the point marked (102) in the figure is representative of (102) planes and the distance of points (102) in the figure from the origin represents $\dfrac{1}{d_{102}}$ and the direction from the origin to (102) intersection point represents the direction of normal to (102) planes. The area enclosed by the reciprocal points 000, 100, 101, 001, represents one face of unit cell of two dimensional reciprocal lattice and as in case of direct lattice we can generate the whole reciprocal lattice by regular repetition of this unit cell.

Rules for finding the reciprocal lattice :

1. From common origin, draw a normal to each plane.

2. Place a point on the normal to each plane (hk$l$) at a distance equal to $\dfrac{1}{d_{hkl}}$ from the origin.

2.5 VECTOR DEVELOPMENT OF RECIPROCAL LATTICE (Relation between the Direct and Reciprocal Lattice)

The relationship between the normal to a plane and the crystallographic axes a, b and c will be developed. This can be obtained by considering a primitive unit cell as shown in Fig. 2.3. Let d_{100} represent height of the cell.

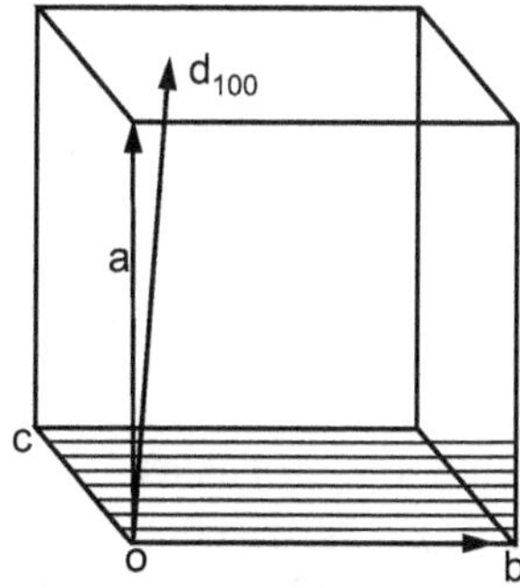

Fig. 2.3

Therefore, volume of unit cell is given by

$$V = \text{Area (shaded base with sides b and c)} \times \text{height}$$

$$= \text{area} \cdot d_{100}$$

$$\therefore \quad \frac{1}{d_{100}} = \frac{\text{area}}{\text{volume}} \quad \quad \dots (2.1)$$

$$\therefore \quad \frac{\vec{n}}{d_{100}} = \frac{\vec{b} \times \vec{c}}{\text{volume}} = \vec{\sigma}_{100}$$

where $\vec{n}$ is the unit normal vector.

$$\therefore \quad \frac{\vec{n}}{d_{100}} = \frac{\vec{b} \times \vec{c}}{\vec{a} \cdot \vec{b} \times \vec{c}}$$

$$\therefore \quad \vec{\sigma}_{100} = \frac{\vec{n}}{d_{100}} = \frac{\vec{b} \times \vec{c}}{\vec{a} \cdot (\vec{b} \times \vec{c})}$$

where, $\vec{a} \cdot (\vec{b} \times \vec{c})$ represents the volume of the primitive unit cell.

Similar expressions can be written for $\vec{\sigma}_{010}$ and $\vec{\sigma}_{001}$. These three vectors are chosen as the three reciprocal axes for defining three dimensional reciprocal lattice. The three reciprocal axes are defined as follows.

$$\left.\begin{array}{l} a^* = \vec{\sigma}_{100} = \dfrac{\vec{b} \times \vec{c}}{\vec{a} \cdot (\vec{b} \times \vec{c})} \\[4ex] b^* = \vec{\sigma}_{010} = \dfrac{\vec{c} \times \vec{a}}{\vec{a} \cdot (\vec{b} \times \vec{c})} \\[4ex] c^* = \vec{\sigma}_{001} = \dfrac{\vec{a} \times \vec{b}}{\vec{a} \cdot (\vec{b} \times \vec{c})} \end{array}\right\} \qquad \dots (2.2)$$

Equation (2.2) gives the reciprocal lattice vector in terms of direct lattice vectors. These three vectors are chosen on three reciprocal axes for defining three dimensional reciprocal lattice. The reciprocal lattice provides simple relationship to the crystal axes.

$$a^* \text{ is normal to } b \text{ and } c$$

$$b^* \text{ is normal to } c \text{ and } a$$

$$c^* \text{ is normal to } a \text{ and } b$$

Applying these conditions, we have

$$a^* \cdot b = 0 = a^* \cdot c$$

$$b^* \cdot c = 0 = b^* \cdot a$$

$$c^* \cdot a = 0 = c^* \cdot b$$

Again it can be proved that

$$
\left.
\begin{aligned}
a^* \cdot a &= \frac{a \cdot b \times c}{a \cdot b \times c} = 1 \\[2mm]
b^* \cdot b &= \frac{a \cdot b \times c}{a \cdot b \times c} = 1 \\[2mm]
c^* \cdot c &= \frac{a \cdot b \times c}{a \cdot b \times c} = 1
\end{aligned}
\right\} \qquad \ldots (2.3)
$$

2.6 PROPERTIES OF RECIPROCAL LATTICE

The characteristic properties of reciprocal lattice are as follows :

(1) Every crystal structure has two lattices associated with it, the crystal lattice and the reciprocal lattice. A diffraction pattern of the crystal is a map of reciprocal lattice of the crystal.

(2) Vectors in the direct lattice have the dimension of length. Vectors in the reciprocal lattice have the dimension of length^{-1}.

(3) A crystal lattice is a lattice in real space. The reciprocal lattice in the Fourier space is associated with the crystal.

(4) Each point in a reciprocal lattice corresponds to a particular set of parallel planes of the direct lattice.

(5) The distance of reciprocal lattice point from an arbitrarily fixed origin is inversely proportional to the interplanar spacing of the corresponding parallel planes of the direct lattice.

(6) The volume of a unit cell of the reciprocal lattice is inversely proportional to the volume of the unit cell of the direct lattice.

(7) The unit cell of the reciprocal lattice need not be a parallelopiped. This is called a Brillouin zone.

(8) The direct lattice is the reciprocal of its own reciprocal lattice.

(9) Simple cubic lattice is its own reciprocal. bcc and fcc lattices are reciprocal to each other.

(a) Show that reciprocal of the reciprocal lattice is the direct lattice :

In order to prove this property, the reciprocal vector (a*) may be written as

$$
(a^*)^* = \frac{b^* \times c^*}{a^* \cdot b^* \times c^*} \qquad \ldots (2.4)
$$

Multiplying R.H.S. of equation (2.4) by $a \cdot a^*$, we get

$$(a^*)^* = a \cdot a^* \frac{b^* \times c^*}{a^* \cdot b^* \times c^*} = a \frac{a^* \cdot b^* \times c^*}{a^* \cdot b^* \times c^*}$$

$$= a \qquad \qquad \dots (2.5)$$

Similarly, we can show

$$(b^*)^* = b \quad \text{and} \quad (c^*)^* = c$$

From equations (2.4) and (2.5), we get

$$a = \frac{b^* \times c^*}{a^* \cdot b^* \times c^*}$$

Similarly, $\qquad b = \dfrac{c^* \times a^*}{a^* \cdot b^* \times c^*}$

and $\qquad c = \dfrac{a^* \times b^*}{a^* \cdot b^* \times c^*}$

(b) To prove the volume of unit cell of the reciprocal lattice is inversely proportional to the volume of unit cell of the direct lattice :

We know that the volume of the direct lattice is $(a \cdot b \times c)$ and volume of the reciprocal lattice is $(a^* \cdot b^* \times c^*)$. We have to show that,

$$a \cdot b \times c = \frac{1}{a^* \cdot b^* \times c^*}$$

We know that,

$$a^* \cdot b^* \times c^* = \left\{ \frac{b \times c}{a \cdot b \times c} \right\} \left\{ \frac{c \times a}{a \cdot b \times c} \right\} \left\{ \frac{a \times b}{a \cdot b \times c} \right\}$$

$$= \frac{1}{[a \cdot b \times c]^3} [(b \times c) \cdot \{(c \times a) \times (a \times b)\}]$$

$$= \frac{1}{[a \cdot b \times c]^3} (b \times c) [\{c \cdot (a \times b)\} a - (a \times b)\} c]$$

$$\qquad \qquad \dots (2.6)$$

But $\quad a \cdot (a \times b) = 0$, hence

$$(b \times c) \{c (a \times b)\} a = [c \cdot (a \times b)] [a \cdot (a \times c)]$$

$$= [a \cdot (b \times c)]^2 \qquad \qquad \dots (2.7)$$

Using equation (2.7) in (2.6), we get

$$a^* \cdot b^* \times c^* = \frac{[a \cdot (b \times c)]^2}{[a \ (b \times c)]^3} = \frac{1}{[a \cdot (b \times c)]}$$

$$\therefore \qquad a \cdot (b \times c) = \frac{1}{a^* \cdot (b^* \times c^*)}$$

(c) Show that every reciprocal lattice vector is perpendicular to a direct lattice plane :

In order to prove that σ_{hkl} is normal to the crystal plane (hk$l$), we have to show that the scalar product of σ_{hkl} and vector lying in the plane (hk$l$) vanish.

The plane (hk$l$) is as shown in Fig. 2.4.

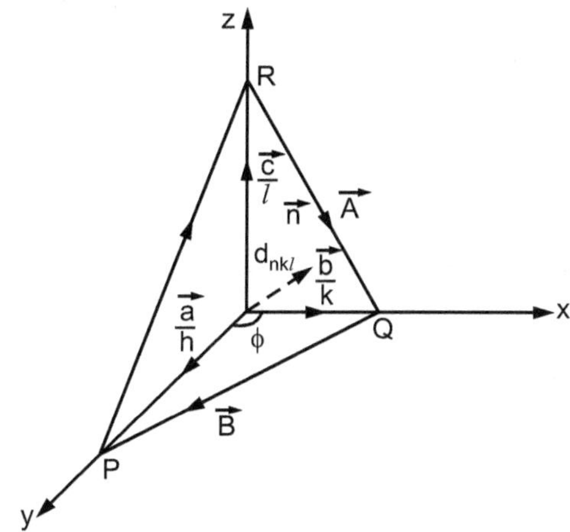

Fig. 2.4

The intercept on x-axis is

$$\vec{OP} = \frac{\vec{a}}{h}$$

The intercept on y-axis is

$$\vec{OQ} = \frac{\vec{b}}{k}$$

The intercept on z-axis is

$$\vec{OR} = \frac{\vec{c}}{l}$$

From geometry of the figure, we have

$$\frac{\vec{b}}{k} + \vec{B} = \frac{\vec{a}}{h}$$

$$\therefore \quad \vec{B} = \frac{\vec{a}}{h} - \frac{\vec{b}}{k} \qquad \qquad \dots (2.8)$$

The scalar product of $\vec{B}$ and $\vec{\sigma}_{hkl}$ is given by

$$\vec{B} \cdot \vec{\sigma}_{hkl} = \left\{ \frac{\vec{a}}{h} - \frac{\vec{b}}{k} \right\} (ha^* + kb^* + lc^*)$$

$$= \frac{a^*}{h}(ha^* + kb^* + lc^*) - \frac{\vec{b}}{k}(ha^* + kb^* + lc^*)$$

$$= \left(\frac{h}{h} + 0 + 0 \right) - \left(0 + \frac{k}{k} + 0 \right)$$

$$= 1 - 1$$

$$= 0 \qquad \qquad \dots (2.9)$$

Also from Fig. 2.4, we can write,

$$\frac{c^*}{l} + \vec{A} = \frac{\vec{b}}{k}$$

$$\therefore \quad \vec{A} = \frac{\vec{b}}{k} - \frac{\vec{c}}{l}$$

$$\therefore \quad \vec{A} \cdot \vec{\sigma}_{hkl} = \left[\frac{\vec{b}}{k} - \frac{\vec{c}}{l} \right] [ha^* + kb^* + lc^*]$$

$$= \left[0 + \frac{k}{k} + 0 \right] - \left(0 + 0 + \frac{l}{l} \right)$$

$$= 1 - 1$$

$$\vec{A} \cdot \vec{\sigma}_{hkl} = 0 \qquad \qquad \dots (2.10)$$

From equations (2.9) and (2.10), we conclude that $\vec{\sigma}_{hkl}$ is normal to $\vec{B}$ and $\vec{A}$. Hence $\vec{\sigma}_{hkl}$ is normal to the plane containing $\vec{B}$ and $\vec{A}$ i.e. to the pane (hkl). Thus every reciprocal lattice vector $\vec{\sigma}_{hkl}$ is perpendicular to the direct lattice plane (hkl).

(d) Prove that the spacing (d_{hkl}) of the plane crystal lattice is equal to $\left|\dfrac{1}{\sigma_{hkl}}\right|$:

Here it is clear that the unit vector $\vec{n}$ is normal to the plane (hkl) and parallel to $\vec{\sigma}_{hkl}$. We can write

$$\vec{n}\,|\sigma_{hkl}| = ha^* + kb^* + lc^*$$

$$\therefore \qquad \vec{n} = \frac{ha^* + kb^* + lc^*}{|\sigma_{hkl}|}$$

The length of the interplaner spacing of the plane is given by

$$d_{hkl} = \frac{\vec{a}}{h}\cos\phi$$

$$= \frac{\vec{a}}{h}\,\vec{n}$$

$$= \frac{\vec{a}}{h}\,\frac{ha^* + kb^* + lc^*}{|\vec{\sigma}_{hkl}|}$$

$$= \frac{1}{|\vec{\sigma}_{hkl}|}\left[\frac{h}{h}\cdot 1 + 0 + 0\right]$$

$$d_{hkl} = \frac{1}{|\vec{\sigma}_{hkl}|}$$

2.7 BRAGG'S LAW IN RECIPROCAL LATTICE [EWALD'S CONSTRUCTION]

Bragg's law in direct lattice is given by $2d_{hkl}\sin\theta = \lambda$, for first order reflection, where d_{hkl} is the interplaner spacing, θ is the glancing angle and λ is the wavelength of incident beam (X-ray). Bragg's law can be represented much more conveniently in terms of reciprocal lattice. Reciprocal lattice helps to understand the wave mechanical behaviour of electron in periodic crystal lattice. Bragg's diffraction condition can be written as

$$\sin\theta_{hkl} = \frac{\lambda/2}{d_{hkl}} = \frac{\lambda/d_{hkl}}{2}$$

A geometrical construction of the reciprocal lattice point is as shown in Fig. 2.5 (a).

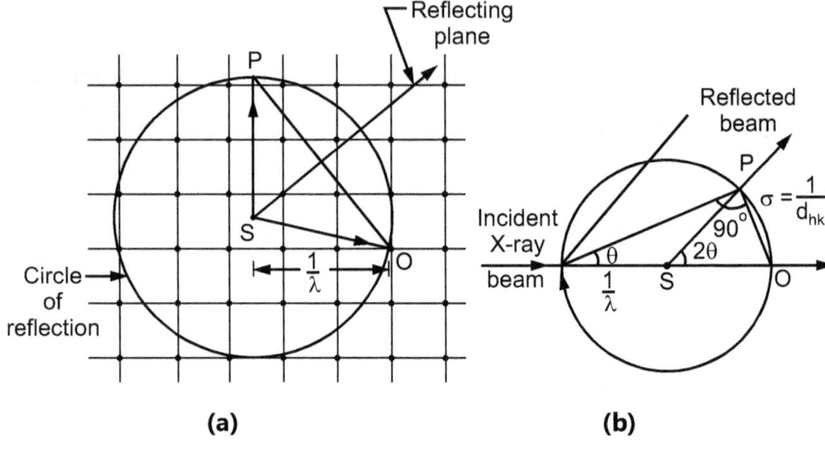

Fig. 2.5

In Fig. 2.5 (a), 'SO' is a vector whose length is $\frac{1}{\lambda}$. This vector is drawn in the direction of incident X-ray beam and ending at the origin of the reciprocal lattice. Now sphere of radius $\frac{1}{\lambda}$ is constructed at a point S as centre. Let this sphere intersect some point (h', k', l') of the reciprocal lattice at P. The vector OP then represents a vector counting the origin of the reciprocal lattice and a point (h' k' l') of that lattice. The vector is normal to (hkl) plane of direct lattice and its length is $\frac{1}{d_{hkl}}$. From Fig. 2.5 (a) the length of the vector OP can be calculated which is $2 \sin \frac{\theta}{\lambda}$. Hence

$$2 \sin \frac{\theta}{\lambda} = \frac{1}{d_{hkl}}$$

$$\therefore \qquad \lambda = 2 d_{hkl} \sin \theta$$

Thus Bragg's condition is satisfied. Thus the vector OP represents a normal to the reflecting plane (hkl) and the vector SP is in the direction of the diffracted beam.

The direction of the diffracted beam is as shown in Fig. 2.5 (b). For any experimental set up, the direction of X-ray beam is defined as AO. Diffraction occurs only when the orientation of the crystal is such that a reciprocal lattice point P comes to lie on the circumference of a circle S of radius $\frac{1}{\lambda}$. When this occurs, a diffracted beam is developed in the direction SP. Thus we have the result that the lattice planes are correctly set for Bragg's diffraction when their reciprocal lattice point lie on the circle of reflection and the direction of diffraction is given by the line joining the origin to the reciprocal lattice point. Because of this property the circle is called as circle of reflection. In three dimensions, the circle is replaced by sphere. This construction is due to Ewald and the sphere of reflection is commonly referred as Ewald sphere.

2.8 POWDER METHOD OF X-RAY DIFFRACTION AND ANALYSIS OF CUBIC CRYSTAL STRUCTURE

The powder method is the only method which can be used with that large class of substance which can be obtained easily in the form of perfect crystals of appreciable size. This method was devised independently by Debye and Scherrer in Germany and by Hill in America.

In this method, a monochromatic X-ray beam is used and fine powders or crystalline aggregates of all kinds having random or chaotic orientation are used. Such a powder requires no rotation because every atomic plane is present in every possible orientation. The entirely random orientation of the grains w.r.t. to the beam means that some of them will be in a position to reflect the radiation from an important set of planes. The diffracted rays go out from individual crystallites which happen to be oriented with planes making an angle θ, with the beam satisfying the Bragg's equation ($2d_{hkl} \sin \theta = n\lambda$), while another fraction of the grains will have another set of planes in the correct position for the reflection and so on. Further reflections are possible not only from the different set of planes but also in different orders for such set, since all the orientations of the fragments are equally likely the reflected rays will form a cone whose axis is in the direction of incident beam and whose semi-vertical angle is 2θ. There is such a cone of diffracted rays for each set of planes. The cones intercept the film in a series of concentric circular halves from the radii of which the angle θ and hence the spacing of the planes can be deduced.

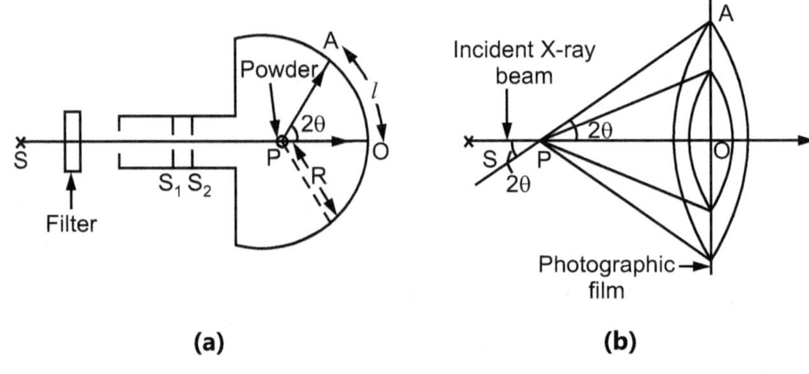

Fig. 2.6

The formation of powder photograph is as shown in Fig. 2.6 (a). The radiation is made monochromatic with the help of filter and it is collimated by passing through the fine slits S_1 and S_2 as shown in Fig. 2.6 (a). Let P be the powder and O is the point where the direct beam would have struck the film. The point A on the film corresponds at which a spectrum with glancing angle θ is formed. The diffracted maxima lie on cones co-axial with direct beam and if a photographic plate is mounted normal to the direct beam, concentric circles are registered upon it as shown in Fig. 2.6 (b). In similar way the diffracted beams from other sets of lattice planes with different interplaner spacing would lie along different cones with different semivertical angles. These cones would be co-axial since the incident beam direction remains the same for all these lattice planes. The photographic film in the form of cylindrical shape is employed whose axis is perpendicular to the beam. There appear arcs of circles on the film on developing. The traces obtained on developed photographic film are shown in Fig. 2.6 (c).

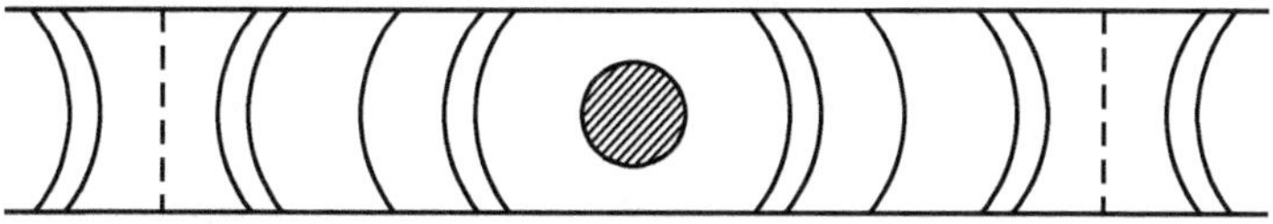

$$\theta = 90 \qquad\qquad \theta = 0$$

Fig. 2.6 (c)

From Fig. 2.6 (c) it is observed that when the rays diffracted through small angles they make arcs around the central spot 'O' on the film. When the rays are diffracted through 90° the cone comes flat and corresponding traces are straight lines. When the diffracted angle is above 90° the curvature is reversed and when the angle approaches to 180° the traces are nearly circular. Thus the curvature of lines changes from the centre to the outside of the film.

Let l be the distance from O to A measured on the film and R be the radius of the camera [Fig. 2.6 (a)], then we have

$$2\theta = \frac{arc}{radius} = \frac{l}{R}$$

$$\therefore \qquad \theta = \frac{l}{2R}$$

In this way be measuring l, the value of θ can be calculated and then interplaner spacing d can be calculated by Bragg's relation ($2d \sin \theta = n\lambda$).

This method is very useful in investigating the structure of simple crystals particularly belonging to the cubic system of which spacing a, b, c of unit cells are all equal. In these crystals there are certain definite relationship between the angles at which the spectra can occur. The spacing of all the planes parallel to faces of the same form (hk$l$) are equal and therefore produce spectra of same angles.

Diffraction Intensities :

The way in which the crystal structure determines the relative intensities of the diffracted beam can be seen by considering the NaCl structure. This was first done by W. L. Bragg, who measured the intensities diffracted by successive orders of the planes (100), (110) and (111) as shown in Fig. 2.7.

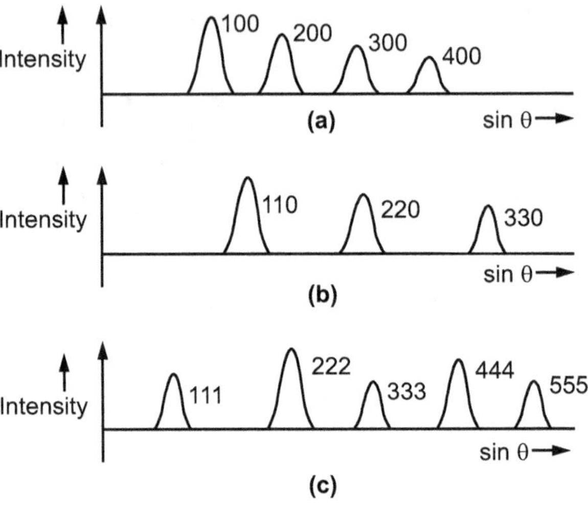

Fig. 2.7

It is noted that the diffraction of successive order n can be pictured as the diffraction by fictitious planes (nh, nk, nl). Thus

$$\lambda = \frac{2d}{n} \sin\theta = 2d_{nh,\, nk,\, nl} \sin\theta$$

Since $\qquad d_{nh,\, nk,\, nl} = \dfrac{d_{hkl}}{n}$

These planes are said to be fictitious because they do not pass through lattice point in the way that the real planes (hk$l$) do. Fig. 2.7 shows observed intensities of the diffracted beams. The significance of the observed intensity variation is that successive orders of (111) reflections are alternately weak and strong. Refer Fig. 2.7 (c). This suggests that (111) planes are composed of inter-leaved layers of unlike atoms because the atoms of one layered set diffract in phase with each other but not with atoms of the other set. Thus in NaCl crystal, alternate (111) planes are composed exclusively of Na and Cl atoms. It is noted that alternate (100) and (110) planes are composed of similar arrays of Na and Cl atoms so that an analogous intensity variation should not occur.

The X-ray scattering power of an atom is directly proportional to the number of electrons composing it and can be expressed by scattering factor (f). The scattering power of an atom actually decreases as θ increases. The scattering factor depends upon the structure factor F_{hkl}

for any reflecting planes (hk$l$). This structure factor F_{hkl} for any reflection (hk$l$) plane is given by

$$F_{hkl} = f_1 \exp [2\pi i (hx_1 + ky_1 + lz_1)] \ldots +$$
$$+ f_n \exp [2\pi i (hx_n + ky_n + lz_n)]$$
$$= \sum_{n=1}^{N} f_n \exp \{2\pi i (hx_n + ky_n + lz_n)] \qquad \ldots (2.11)$$

where $x_1 \, y_1 \, z_1 \ldots x_n \, y_n \, z_n$ are position co-ordinates of atoms in the unit cell of the crystal. N is the total number of atoms in the unit cell and n is the function of its position in the unit cell x_n, y_n and z_n. Thus if the distribution of atoms among available equipoints are known then the structure factor can be determined for different (hk$l$) reflections.

The unit cell of NaCl contains four Na atoms at 000, $\frac{1}{2}\frac{1}{2}0$, $\frac{1}{2}0\frac{1}{2}$, $0\frac{1}{2}\frac{1}{2}$ and four Cl atoms at $\frac{1}{2}\frac{1}{2}\frac{1}{2}$, $\frac{1}{2}00$, $0\frac{1}{2}0$, $00\frac{1}{2}$ respectively.

Therefore, the structure factor for NaCl becomes

$$F_{hkl} = f_{Na} \left[\exp \{2\pi i \,(000)\} + \exp \left\{ 2\pi i \left(\frac{h}{2} + \frac{k}{2} + 0 \right) \right\} \right.$$

$$\left. + \exp \left\{ 2\pi i \left(\frac{h}{2} + 0 + \frac{l}{2} \right) \right\} + \exp \left\{ 2\pi i \left(0 + \frac{k}{2} + \frac{l}{2} \right) \right\} \right]$$

$$+ f_{Cl} \left[\exp \left\{ 2\pi i \left(\frac{h}{2} + \frac{k}{2} + \frac{l}{2} \right) \right\} + \exp \left\{ 2\pi i \left(\frac{h}{2}, 0, 0 \right) \right\} \right.$$

$$\left. + \exp \left\{ 2\pi i \left(0 + \frac{k}{2} + 0 \right) \right\} + \exp \left\{ 2\pi i \left(0, + 0 + \frac{l}{2} \right) \right\} \right]$$

$$= [f_{Na} + f_{Cl} \, e^{\pi i \,(h + k + l)}]$$
$$[e^0 + e^{\pi i \,(h + k)} + e^{\pi i \,(h + l)} + e^{\pi i \,(k + l)}] \qquad \ldots (2.12)$$

Now it should be noted that

$$e^{\pi i n} = 1, \text{ for even value of n including zero}$$
$$= -1 \text{ for odd value of n}$$

This means that Miller indices (hk$l$) are not all even or odd i.e. they are mixed. The terms in the second parentheses in equation (2.12) equal to zero since two exponential terms are +1 and are −1. When the indices are not mixed, the term in second parentheses add upto four, so that the structure for NaCl can be written as

$$F_{hkl} = 4\left[f_{Na} + f_{Cl}\, e^{\pi i\,(h+k+l)}\right] \qquad \ldots (2.13)$$

This can be further simplified by noting that

$$F_{hkl} = 4\left[f_{Na} + f_{Cl}\right] \text{ when } hkl \text{ are all even}$$

$$F_{hkl} = 4\left[f_{Na} - f_{Cl}\right] \text{ when } hkl \text{ are all odd} \qquad \ldots (2.14)$$

The intensity of X-ray reflected by planes (hkl) in a crystal is actually proportional to the square of their structure factor. The equation (2.14) clearly shows why reflections (111), (333) have a lower intensity than the reflections (222), (444) ... in the NaCl structure. Also we know that the reflections from the planes having mixed indices cannot be observed because their structure factors are identically equal to zero. Such reflections are space-group extinctions and occur in all crystals having centred lattices. Specifically it can be shown that the extinction rule for all face-centred lattice requires that (hkl) be unmixed for a reflection to occur.

Substituting the values of $\sin\theta$ for reflection from (100), (110) and (111) planes, the Bragg's relation becomes

$$\left.\begin{array}{l} n_1\lambda = 2d_{100}\,(0.126) \\[2mm] n_2\lambda = 2d_{110}\,(0.178) \\[2mm] n_3\lambda = 2d_{111}\,(0.109) \end{array}\right\} \qquad \ldots (2.15)$$

From geometry of the cubic crystal lattice, we have

$$d_{100} = \sqrt{2}\, d_{110}, \quad d_{100} = \sqrt{3}\, d_{111}, \quad d_{110} = \sqrt{\frac{3}{2}}\, d_{111}$$

Using these relations in equation (2.15) and eliminating λ, we have

$$\left.\begin{array}{l} \dfrac{d_{100}}{d_{110}} = \dfrac{(0.178)\,n_1}{(0.126)\,n_2} = \sqrt{2} \\[4mm] \dfrac{d_{100}}{d_{111}} = \dfrac{(0.109)\,n_1}{(0.126)\,n_3} = \sqrt{3} \\[4mm] \dfrac{d_{110}}{d_{111}} = \dfrac{(0.109)\,n_2}{(0.178)\,n_3} = \sqrt{\dfrac{3}{2}} \end{array}\right\} \qquad \ldots (2.16)$$

It is clearly seen that the last two relations in equation (2.16) can be satisfied only if $n_1 = 2$, $n_2 = 2$ and $n_3 = 1$. Thus it is clear that the indices shown in upper two graphs in Fig. 2.7 should be doubled. This supports the conclusion previously reached that the indices of allowed reflections from NaCl structure must be all odd or even.

Determination of unit cell contents :

It is not possible to determine the interplanar distance from equation (2.16), since there are three unknown values and only two independent equations relating them. Therefore it is not possible to calculate their absolute values. To calculate this the use of the relation between unit cell volume (V) and density (ρ) of NaCl crystal is done.

$$\therefore \quad \text{Density } (\rho) = \frac{\text{Mass of the unit cell}}{\text{Volume of the unit cell}}$$

$$= \frac{nm/N_o}{V}$$

where,

n = number of molecules per unit cell

m = atomic weight in atomic mass unit

N_o = Avogadro's number

$(6.023 \times 10^{26}$ molecules/kg-mole$)$

V = volume of unit cell in m^3

The density of NaCl is 2.15 gm/cc = 2150 kg/m^3.

For NaCl crystal the number of molecules per unit cell is

$$(n) = \frac{1}{8} \times 8 + \frac{1}{2} \times 6 = 4.$$

Molecular weight of NaCl = 35.5 + 23 = 58.5 amu.

The unit cell volume is 178×10^{-27} m^3. The unit cell of NaCl is cubic, so that the length of its unit cell edge is

$$a = V^{1/3} = 5.63 \times 10^{-10} \text{ m}$$

Obviously, the distance between two adjacent ions Na and Cl is,

$$d = \frac{a}{2} = 2.813 \times 10^{-10} \text{ m}$$

Since a = d_{100} = $2d_{100}$ in cubic crystals, it is now possible to determine the absolute length of the interplaner spacing. Substituting these values in Bragg's equation, it is possible to determine the wavelength of X-ray radiation used to measure reflection intensities. When X-ray wavelength is known, it is possible to determine (d) from the measured reflection angle (θ). Thus the knowledge of interplaner spacing values permits the determination of the lattice type and the unit cell volume.

SOLVED PROBLEMS

Problem 2.1 : At what angle may ray of wavelength 0.440 A° be reflected from a face of a rock salt crystal [d = 2.814 A°] ?

Solution : We have,

$$2d \sin \theta = n\lambda$$

$$\therefore \qquad \sin \theta = \frac{n\lambda}{2d}$$

For 1st order, n = 1,

$$\sin \theta_1 = \frac{1 \cdot \lambda}{2d} = \frac{1 \times 0.440}{2 \times 2.814} = 0.0782$$

$$\therefore \qquad \theta_1 = \mathbf{4°, \ 29'}$$

For 2nd order, n = 2,

$$\sin \theta_2 = \frac{2\lambda}{2d} = \frac{2 \times 0.440}{2 \times 2.814} = 2 \times 0.0782$$

$$\therefore \qquad \theta_2 = \mathbf{8°, \ 59'}$$

For 3rd order, n = 3,

$$\sin \theta_3 = 3 \times 0.0782$$

$$\therefore \qquad \theta_3 = \mathbf{13°, \ 34'}$$

Hence the reflected beams will be observed at angles 4°, 29'; 8°, 59' and 13°, 34' etc.

Problem 2.2 : When X-ray beam is incident on NaCl crystal with lattice spacing 2.82×10^{-10} m, the first order Bragg's reflection is observed at a glancing angle of 8°, 35'. What is the wavelength of X-rays ? At what angle would the second order Bragg's reflection occur ?

Solution : According to Bragg's law, we have

$$2d \sin \theta = n\lambda$$

$$\therefore \quad \lambda = \frac{2d \sin \theta}{n} = \frac{2 \times 2.82 \times 10^{-10} \cdot \sin (8°, 35')}{1}$$

$$= \mathbf{0.842 \ A°}$$

For second order, n = 2

$$\therefore \quad \sin \theta = \frac{2\lambda}{2d} = \frac{\lambda}{d} = \frac{0.842 \times 10^{-10}}{2.82 \times 10^{-10}} = 0.2986$$

$$\theta = \mathbf{17°, \ 22'}$$

Problem 2.3 : The Bragg's angle corresponding to the first order reflection from (1, 1, 1) planes in a crystal is 30° when X-rays of wavelength 1.75 A° are used. Calculate the interatomic spacing.

Solution : We know that,

$$d_{hkl} = \frac{a}{\sqrt{h^2 + k^2 + l^2}}$$

According to this problem, h = k = l = 1, θ = 30°

$$\lambda = 1.75 \times 10^{-10} \text{ m and n = 1.}$$

We have, $2d \sin \theta = n\lambda$

$$2 \times \frac{a}{\sqrt{3}} \cdot \sin 30 = 1 \times 1.75 \times 10^{-10} \qquad \left[\text{since } d_m = \frac{a}{\sqrt{3}} \right]$$

$$\therefore \quad a = 1.732 \times 1.75 \times 10^{-10}$$

$$a = \mathbf{3.031 \ A°}$$

Problem 2.4 : Calculate the glancing angle on the cube (100) of rock salt crystal (a = 2.814 A°) corresponding to second order diffraction for X-rays of wavelength 0.710 A°.

Solution : According to Bragg's law, we have

$$2d \sin \theta = n\lambda$$

The distance between consecutive lattice planes is

$$d_{hkl} = \frac{a}{\sqrt{h^2 + k^2 + l^2}}$$

$$d_{100} = \frac{2.814}{\sqrt{1^2 + 0 + 0}} = 2.814 \ A°$$

$$\therefore \quad 2 \times 2.814 \sin \theta = 2 \times 0.710$$

$$\therefore \quad \theta = \textbf{14°, 36'}$$

EXERCISES

(A) Multiple Choice Type Questions :

1. Diffraction of X-rays from the crystal is the phenomenon of ......

   (a) scattering   **(b) reflection**

   (c) refraction   (d) interference

2. The reciprocal of the reciprocal lattice is ...... lattice.

   (a) direct   (b) indirect

   (c) positive   (d) negative

3. The reciprocal lattice vector is ...... to a direct lattice plane.

   (a) parallel   **(b) perpendicular**

   (c) addition   (d) none of these

4. The powder method of X-ray diffraction is very useful to investigate the crystal structure particularly belonging to ...... system.

   (a) tetragonal crystal   **(b) cubic crystal**

   (c) hexagonal crystal   (d) octahedral crystal

5. The Millar indices of the plane parallel to Y and Z axes are ......

   (a) (001)   (b) (010)

   (c) (1, 0, 0)   (d) (111)

6. The volume of unit cell of the reciprocal lattice is ......

    (a) directly proportional to the volume of unit cell of the direct lattice

    (b) inversely proportional to the volume of unit cell of the direct lattice

    (c) equal to the volume of the primitive cell

    (d) not equal to the volume of the primitive cell

7. The number of molecules present in the unit cell of NaCl is ......

    (a) 2   (b) 5

    (c) 4   (d) none of these

8. If X-rays are reflected from two parallel planes, one above the other, it satisfies the condition of path difference as ......

    (a) $2d \sin \theta = n\lambda$ (b) $d \sin \theta = n\lambda$

    (c) $\sin \theta = n\lambda$ (d) $2d \cos \theta = n\lambda$

(B) Short Answer Type Questions :

    1. Explain the concept of reciprocal lattice.

    2. Obtain the relation between direct and reciprocal lattice.

    3. Show that the magnitude of the reciprocal lattice vector is equal to $\left| \dfrac{d}{d_{hkl}} \right|$.

    4. Give geometrical demonstration of reciprocal lattice in two dimensions.

    5. State the properties of reciprocal lattice.

(C) Long Answer Type Questions :

    1. Discuss Ewald construction of Bragg's law of X-ray diffraction in the reciprocal lattice.

    2. Explain with neat diagram of powder method of X-ray diffraction.

3. With the help of powder method, discuss relative intensities of diffracted beam.

4. Give geometrical construction of the reciprocal lattice.

5. Explain the properties of the reciprocal lattice.

6. Show that every reciprocal lattice vector is normal to the lattice plane of the crystal lattice.

7. Explain the concept of reciprocal lattice. Hence show that the reciprocal of the reciprocal lattice is the direct lattice.

(D) Problems for Practice :

1. In NaCl crystal (a = 2.814 A°) the second order reflection is observed from the plane (100) by using monochromatic X-ray beam of wavelength 0.71 A°. Calculate the angle of diffraction to observe the above reflection.

$$\left[\textbf{Hint : } d_{hkl} = \frac{1}{\sqrt{h^2 + k^2 + l^2}} \right]$$
    (**Ans.** θ = 14°, 61')

2. A powder pattern is obtained for lead with radiations of λ = 1.54 A°. The (220) reflection is observed at Bragg's angle θ = 32°. What is the lattice parameter of the atom ?

    (**Ans.** a = 4.1×10^{-10} m)

3. The primitive translation vectors of the horizontal space lattice are $\vec{a} = \frac{a}{2}x + \frac{\sqrt{3}\,a}{2}y$, $\vec{b} = -\frac{a}{2}x + \frac{\sqrt{3}\,a}{2}y$ and $\vec{c} = cz$. Calculate the volume of the primitive cell.

    $$\left(\textbf{Ans. } \frac{\sqrt{3}}{2}\,a^2 c \right)$$

    [**Hint :** Find $\vec{a} \cdot \vec{b} \times \vec{c}$]

3
CHAPTER

LATTICE VIBRATIONS

SYLLABUS

Elastic vibration of linear one-dimensional mono-atomic lattice, Expression for frequency and dispersion curve, Elastic vibration of linear one dimensional diatomic lattice – optical and acoustical excitations in ionic crystals, Experimental determination of dispersion relations.

3.1 INTRODUCTION

A space lattice is regarded as the array of atoms or unit cell in three dimensions. By the term lattice vibration, we mean with the vibration of atoms about their mean position in a solid. It has been assumed that the atoms are connected with each other by elastic springs. When an external force is applied to excite any atom, the motion of this atom is shared by all other atoms in crystal and a crystal vibrates as a whole. These are called as thermal vibrations of the crystal or lattice vibration. Such a vibration takes place at any temperature. Hence they are responsible for thermal properties of solid and give contribution to the heat capacity of metal. It has been assumed that the lattices are not continuous and homogeneous but they are composed of discrete atoms.

The aim of this chapter is to study the characteristics of crystal vibrations in one dimensional mono-atomic and diatomic linear lattices. For simplicity, we will consider only one dimensional lattice which may be extended to two and three dimensional lattices.

3.2 ELASTIC VIBRATIONS OF LINEAR ONE DIMENSIONAL MONO-ATOMIC LATTICE

Let us consider the mono-atomic linear lattice in one dimension extending along x-axis as shown in Fig. 3.1.

Let us assume that the distance between the nearest neighbours is 'a' i.e. interatomic distance and each has mass 'm'. It is also assumed that during vibration of atoms there exists interaction only in the neighbouring atoms and Hooke's law is obeyed.

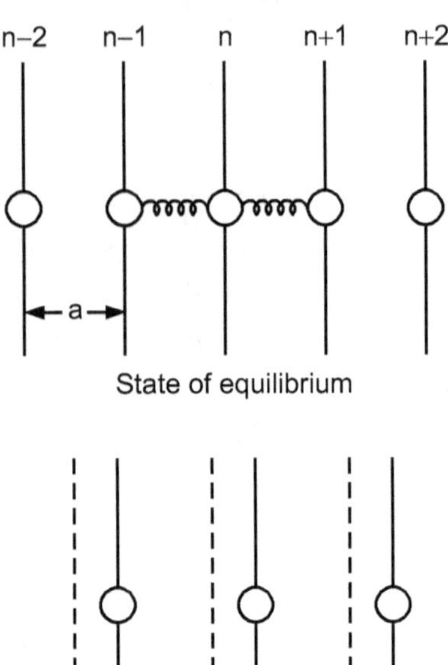

Fig. 3.1

For simplicity we consider the propagation along x-direction and will discuss only longitudinal mode. Under the equilibrium condition the atoms are equally spaced and when the vibrational motion is excited then they will execute simple harmonic motion about their mean position.

Let x_0, x_1, x_2 ... x_{n-1}, x_n, x_{n+1} ... etc. represent the displacement of atoms from equilibrium position.

In order to set up the equation of force, we assume the force of interaction between nearest neighbours. The force of interaction acting on n^{th} atoms can be calculated in terms of the extension of two springs which binds it with atoms (n + 1) and (n – 1), hence the equation of motion of nth particle is given by

$$m\ddot{x}_n = f\,[x_{n+1} - x_n] - f\,[x_n - x_{n-1}]$$

$$= f\,[x_{n+1} + x_{n-1} - 2x_n] \qquad \ldots (3.1)$$

where f is force constant or force of interaction per unit displacement.

According to Newton's law, we can write,

$$m\,\frac{d^2 x_n}{dt^2} = f\,[x_{n+1} + x_{n-1} - 2x_n] \qquad \ldots (3.2)$$

Let the solution of this equation is

$$x_n = A \cdot e^{-i(\omega t - kna)} \qquad \ldots (3.3)$$

where $\qquad A$ = amplitude of the particle

$$k = \frac{2\pi}{\lambda} = \text{wave vector}$$

Differentiating equation (3.3) twice w.r.t. 't', we get,

$$\frac{dx_n}{dt} = A(-i\omega)e^{-i(\omega t - kna)}$$

$$\frac{d^2x_n}{dt^2} = A(-i\omega)^2 e^{-i(\omega t - kna)}$$

$$\ddot{x}_n = -A\omega^2 e^{-i(\omega t - kna)} \qquad \ldots (3.4)$$

Again from equation (3.3), we can write

$$x_{n+1} = Ae^{-i[\omega t - k(n+1)a]} \qquad \ldots (3.5)$$

$$x_{n-1} = Ae^{-i[\omega t - k(n-1)a]} \qquad \ldots (3.6)$$

Substituting equatons (3.4), (3.5) and (3.6) in equation (3.2), we get

$$-mA\omega^2 e^{-i[\omega t - kna]} = fA\left[e^{-i\{\omega t - k(n-1)a\}} + e^{-i\{\omega t + k(n+1)a\}}\right.$$

$$\left. - 2e^{-i\{\omega t - kna\}}\right]$$

$$-m\omega^2 = f\left[e^{-ika} + e^{ika} - 2\right]$$

$$= f\left[2\left(\frac{e^{-ika} + e^{ika}}{2}\right) - 2\right]$$

$$= 2f\left[\cos ka - 1\right]$$

$$\therefore \quad \frac{m\omega^2}{2f} = 1 - \cos ka$$

$$\therefore \quad -m\omega^2 = 2f\left[-2\sin^2\frac{ka}{2}\right]$$

$$= -4f\sin^2\left(\frac{ka}{2}\right)$$

$$\therefore \quad \omega^2 = \frac{4f}{m}\sin^2\left(\frac{ka}{2}\right)$$

$$\therefore \quad \omega = \sqrt{\frac{4f}{m}} \cdot \sin\left(\frac{ka}{2}\right) \qquad \ldots (3.7)$$

$$\text{or} \quad \omega = \omega_{max} \sin\left(\frac{ka}{2}\right)$$

$$\text{where} \quad \omega_{max} = \sqrt{\frac{4f}{m}}$$

The equation (3.7) is called as dispersion relation since ω depends upon the value of $\sin\left(\frac{ka}{2}\right)$.

3.3 EXPRESSION FOR FREQUENCY AND DISPERSION CURVE

The graph of ω against k for lattice with interaction only between nearest neighbour planes is as shown in Fig. 3.2.

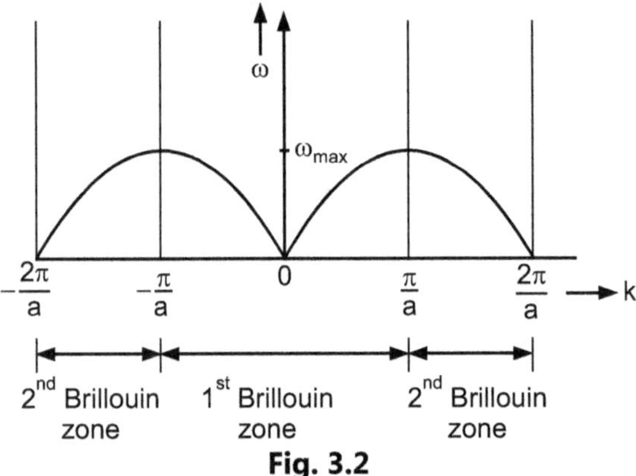

Fig. 3.2

From Fig. 3.2 it is clear that the dispersion curve is periodic function of k with period $\frac{2\pi}{a}$.

The maximum frequency corresponding to the values of k can be obtained from equation (3.7).

$$\omega = 2\pi\upsilon = 2\sqrt{\frac{f}{m}} \cdot \sin\left(\frac{ka}{2}\right)$$

$$\therefore \quad \upsilon = \frac{1}{\pi}\sqrt{\frac{f}{m}} \cdot \sin\left(\frac{ka}{2}\right)$$

$$\therefore \quad \upsilon = \upsilon_{max} = \frac{1}{\pi}\sqrt{\frac{f}{m}} \qquad \ldots (3.8)$$

Thus there exists a maximum frequency υ_{max} which can be propagated through the array of atoms in the crystal. The chain of linear lattice may be treated as low-pass filter which transmits only frequencies between zero and υ_{max}. There is no such frequency limit in case of continuous and homogeneous elastic medium. Also the velocity of propagation of wave becomes smaller as the wavelength decreases in case of crystalline medium, whereas it remains constant in case of continuous medium. The value of frequency of the chain atoms will be maximum when the wave vector is $k = \dfrac{2\pi}{a}$.

Results : [Group velocity (v_g) and phase velocity (v_p)].

(i)   At low frequency ($k \to 0$ the long wavelength limit), if $k << \dfrac{\pi}{a}$, then ω will be approximately linear with k.

$$\therefore \qquad ka << \pi \quad \text{i.e.} \quad \frac{ka}{2} << \frac{\pi}{2}$$

$$\therefore \qquad \sin\frac{ka}{2} \approx \frac{ka}{2}$$

From equation (3.7), we get

$$\omega = 2\sqrt{\frac{f}{m}} \cdot \frac{ka}{2} = \sqrt{\frac{f}{m}} \cdot ka$$

$\therefore \quad \omega \propto k$ for given crystal.

Thus, for long wavelength approximation, the phase velocity will be essentially constant, since

$$\text{Phase velocity,} \qquad v_p = \frac{\omega}{k} = a\sqrt{\frac{f}{m}} = v_o = \text{constant}$$

$$\text{Group velocity,} \qquad v_g = \frac{d\omega}{dk} = a\sqrt{\frac{f}{m}} = v_o = \text{constant}$$

Thus for long wavelength, the dispersion effects are negligible and the medium acts like continuous and homogeneous elastic medium.

(ii)   As k increases (λ decreases) then ω will not be a linear function of k. Under these conditions, phase velocity and group velocity are given by equation (3.7).

$$\omega = 2\sqrt{\frac{f}{m}} \cdot \sin\frac{ka}{2}$$

$$\therefore \quad \frac{\omega}{k} = \frac{2}{k}\sqrt{\frac{f}{m}} \cdot \sin\frac{ka}{2} = \sqrt{\frac{f}{m}} \cdot \frac{\sin\frac{ka}{2}}{\frac{k}{2}}$$

$$= a\sqrt{\frac{f}{m}} \cdot \frac{\sin\frac{ka}{2}}{\frac{ka}{2}}$$

$$= v_o \frac{\sin\frac{ka}{2}}{\frac{ka}{2}}$$

$$\therefore \quad \text{Phase velocity,} \quad v_p = \frac{\omega}{k} = v_o \frac{\sin\frac{ka}{2}}{\frac{ka}{2}}$$

Again we have,
$$\omega = 2\sqrt{\frac{f}{m}} \cdot \sin\frac{ka}{2}$$

$$\therefore \quad d\omega = 2\sqrt{\frac{f}{m}} \cdot \cos\frac{ka}{2} \cdot \frac{a}{2}\, dk$$

$$= a\sqrt{\frac{f}{m}} \cdot \cos\frac{ka}{2}\, dk$$

$$= v_o \cos\frac{ka}{2} \cdot dk$$

$$\therefore \quad \text{Group velocity} = v_g = \frac{d\omega}{dk} = v_o \cos\frac{ka}{2}$$

(iii) If $k = \frac{2\pi}{\lambda} = \pm\frac{\pi}{4}$ i.e. $\lambda = \pm 2a$

$$\therefore \quad v_g = v_o \cos\frac{ka}{2} = v_o \cos\frac{\pi}{a} \cdot \frac{a}{2} = v_o \cos\frac{\pi}{2} = 0$$

This means that the group velocity is zero at the boundaries of first Brillouin zone as shown in Fig. 3.3.

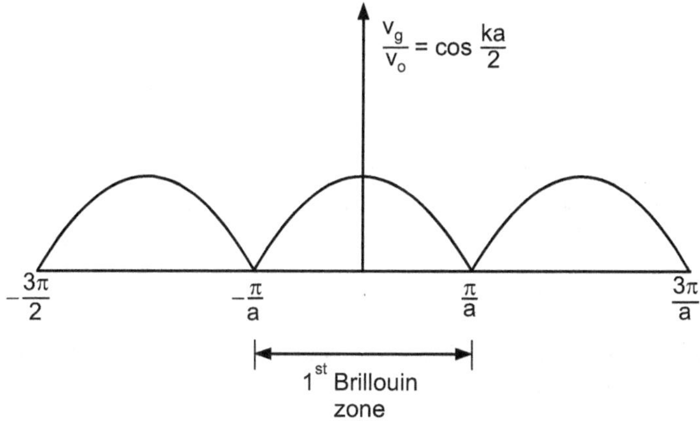

Fig. 3.3

Hence, the boundaries of the first Brillouin zone act just like nodal points in a steady wave.

(iv) **Brillouin zone :** The ratio of the displacement of two successive places from equation (3.3),

$$\frac{x_{n+1}}{x_n} = \frac{Ae^{-i[\omega t - k(n+1)]a}}{Ae^{-i[\omega t - kna]}}$$

$$= e^{ika}$$

This shows that the range $-\pi$ to $+\pi$ for the phase ka covers all independent values of e^{ika}. The range of independent values of k can be specified by

$$-\pi \le ka \le \pi$$

This range represents the first Brillouin zone of linear lattice. The maximum value of this zone is $k_{max} = \pm \dfrac{\pi}{a}$. The second zone is made up of two intervals of half a period each, one on each side of the first Brillouin zone as shown in Fig. 3.3. Similarly, we can represent the higher order zones.

At the boundaries of the first Brillouin zone, the solution given by equation (3.3) does not represent travelling wave but a standing wave, hence the solution given by equation (3.3) becomes

$$x_n = Ae^{-i[\omega t - (\pm n\pi)]}$$

$$= Ae^{-i\omega t} e^{\pm in\pi} = Ae^{i\omega t} \cdot \cos n\pi$$

This is the equation of standing wave in which alternate atoms move in opposite phase, because $\cos n\pi = \pm 1$ according to n is even or odd integer. This condition is equivalent to Bragg's reflection of X-rays where the travelling wave cannot propagate in the lattice. The value of $k_{max} = \pm \dfrac{\pi}{a}$ satisfies the Bragg's condition.

3.4 ELASTIC VIBRATION OF LINEAR ONE DIMENSIONAL DIATOMIC LATTICE

To study this, let us assume a diatomic linear lattice in one dimension as shown in Fig. 3.4.

2n–2   2n–1   2n   2n+1   2n+2

m_2 m_1 m_2 m_1 m_2 m_1

|← a →|← a →|

Fig. 3.4

The dispersion relation shows two branches called as acoustical and optical branches, out of which the optical branch is active in infra-red absorption. The atoms are numbered in such a way that the odd number have mass m_1 and even number have mass m_2. Let 'a' be the distance between nearest neighbours, so that the atoms of the same nature are at a distance 2a part.

Let x_{2n} represent the displacement of $2n^{th}$ particle from the equilibrium position. Similarly, x_{2n-1} and x_{2n+1} denote the displacement of $(2n - 1)^{th}$ and $(2n + 1)^{th}$ particle from equilibrium position respectively.

Considering the nearest neighbours interaction only, so the equation of motion of $2n^{th}$ and $(2n + 1)^{th}$ particle is

$$m_2\ddot{x}_{2n} = f(x_{2n-1} + x_{2n+1} - 2x_{2n}) \qquad \text{... (3.9)}$$

$$m_1\ddot{x}_{2n+1} = f(x_{2n} + x_{2n+2} - 2x_{2n+1}) \qquad \text{... (3.10)}$$

where, f is force constant or force of interaction per unit displacement.

The solution of this equation in the form of running wave with different amplitude is given by

$$x_{2n} = Be^{-i[\omega t - k \cdot 2na]} \qquad \ldots (3.11)$$

$$x_{2n+1} = Ae^{-i[\omega t - q(2n+1)a]} \qquad \ldots (3.12)$$

where $k = \dfrac{2\pi}{\lambda}$ is wave vector and the constants A and B are amplitudes corresponding to atoms of odd mass m_1 and even mass m_2 respectively.

Differentiating equations (3.11) and (3.12) twice w.r.t. 't', we get

$$\frac{dx_{2n}}{dt} = \dot{x}_{2n} = B(-i\omega)e^{-i[\omega t - k2na]}$$

$$\frac{d^2x_{2n}}{dt^2} = \ddot{x}_{2n} = -B(-i\omega)^2 e^{-i[\omega t - k \cdot 2na]}$$

$$= -B\omega^2 e^{-i[\omega t - k \cdot 2na]} \qquad \ldots (3.13)$$

Similarly,

$$\frac{d^2x_{2n+1}}{dt^2} = \ddot{x}_{2n+1} = -A\omega^2 e^{-i[\omega t - k(2n+1)a]} \qquad \ldots (3.14)$$

Substituting the values from equations (3.11) and (3.12) in equation (3.9), we get,

$$-m_2 B\omega^2 e^{-i[\omega t - k \cdot 2na]} = f[Ae^{-i[\omega t - k(2n-1)a} + Ae^{-i[\omega t - k(2n+1)a}$$

$$-2Be^{-i[\omega t - k2na]}]$$

$$-m_2 B\omega^2 e^{-i[\omega t - k \cdot 2na]} = fAe^{-i[\omega t - k(2n-1)a]} + fAe^{-i[\omega t - k(2n+1)a]}$$

$$-2fBe^{-i[\omega t - k2na]}$$

$$\therefore \quad B(m_2\omega^2 - 2f)e^{-i[\omega t - k \cdot 2na]} + fA[e^{ika} + e^{-ika}]e^{-i[\omega t - k \cdot 2na]} = 0$$

$$\therefore \qquad\qquad B(m_2\omega^2 - 2f) + 2Af\cos ka = 0 \qquad \ldots (3.15)$$

Similarly, $\qquad\qquad A(m_1\omega^2 - 2f) + 2Bf\cos ka = 0 \qquad \ldots (3.16)$

If the determinants of the coefficients of A and B are zero then the system has non-vanishing solutions for the constants A and B i.e.

$$\begin{vmatrix} (m_2\omega^2 - 2f) & 2f\cos ka \\ 2f\cos ka & m_1\omega^2 - 2f \end{vmatrix} = 0 \qquad \ldots (3.17)$$

$\therefore \quad (m_2\omega^2 - 2f)(m_1\omega^2 - 2f) - 4f^2 \cos^2 ka = 0$

$m_1 m_2 \omega^2 - m_2 \omega^2 \cdot 2f - m_1 \omega^2 \cdot 2f + 4f^2 - 4f^2 \cos^2 ka = 0$

$m_1 m_2 \omega^2 - (m_1 + m_2)\omega^2 \cdot 2f + 4f^2 [1 - \cos^2 ka] = 0$

$$\therefore \quad \omega^4 - \left[\frac{1}{m_1} + \frac{1}{m_2}\right]\omega^2 \cdot 2f + \frac{4f^2}{m_1 m_2}\sin^2 ka = 0 \qquad \ldots (3.18)$$

This is a quadratic equation in ω^2, whose solution is given by

$$\omega^2 = \frac{\left[\dfrac{1}{m_1} + \dfrac{1}{m_2}\right] 2f \pm \sqrt{\left[\dfrac{1}{m_1} + \dfrac{1}{m_2}\right]^2 4f^2 - 4\dfrac{4f^2}{m_1 m_2}\sin^2 ka}}{2}$$

$$\omega^2 = f\left[\frac{1}{m_1} + \frac{1}{m_2}\right] \pm f\left[\left(\frac{1}{m_1} + \frac{1}{m_2}\right)^2 - \frac{4\sin^2 ka}{m_1 m_2}\right]^{1/2} \qquad \ldots (3.19)$$

The equation (3.19) is called as dispersion relation for diatomic linear lattice. Again from equation (3.19), we get two angular frequencies (ω_-) and (ω_+) for single value of wave vector (k).

Now a graph is plotted between ω and k as shown in Fig. 3.5.

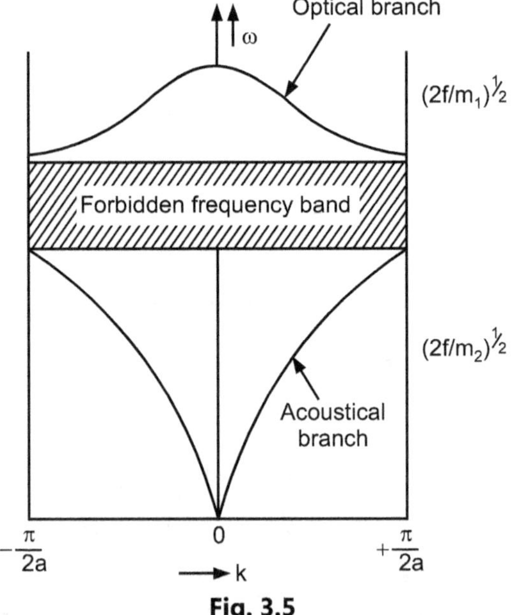

Fig. 3.5

From Fig. 3.5, we get two branches, one corresponding to (ω_-) known as acoustical branch and the other corresponding to (ω_+) known as optical branch.

(a) Optical branch :

For small value of k, i.e. $k \to 0$ and $m_1 > m_2$, we can neglect the sine term in equation (3.19).

$$(\omega_+^2) = f\left[\frac{1}{m_1} + \frac{1}{m_2}\right] + f\left[\frac{1}{m_1} + \frac{1}{m_2}\right]$$

$$= 2f\left[\frac{1}{m_1} + \frac{1}{m_2}\right]$$

$$= 2f\left[\frac{1}{m_2}\right]$$

$$\therefore \quad (\omega_+) = \left(\frac{2f}{m_2}\right)^{1/2} \qquad \qquad \ldots (3.20)$$

This gives optical branch (higher frequency).

(b) Acoustical branch :

Let us consider (ω_-) for $k \to 0$. We neglect the sine term in equation (3.19). Its value is $\dfrac{-4k^2a^2}{m_1m_2}$, $\therefore \sin ka = ka$.

Therefore equation (3.19) becomes,

$$(\omega_-^2) = f\left[\frac{1}{m_1} + \frac{1}{m_2}\right] - f\left[\left(\frac{1}{m_1} + \frac{1}{m_2}\right)^2 - \frac{4k^2a^2}{m_1m_2}\right]^{1/2}$$

$$= f\left[\frac{m_1 + m_2}{m_1m_2}\right] - f\left[\frac{m_1 + m_2}{m_1 + m_2}\right]\left[1 - \frac{4k^2a^2\, m_1m_2}{(m_1 + m_2)^2}\right]^{1/2}$$

$$= f\left[\frac{m_1 + m_2}{m_1m_2}\right]\left[1 - \frac{4k^2a^2\, m_1m_2}{(m_1 + m_2)^2}\right]^{1/2} \qquad \ldots (3.21)$$

But $\left[1 - \dfrac{2k^2a^2\, m_1m_2}{(m_1 + m_2)^2}\right]^2 = 1 + \dfrac{4k^4a^4m_1^2m_2^2}{(m_1 + m_2)^4} - \dfrac{2 \times 2k^2a^2m_1m_2}{(m_1 + m_2)^2}$

$$= 1 - \frac{4k^2a^2m_1m_2}{(m_1 + m_2)^2}$$

Here 2nd term is neglected because k is small.

$$\therefore \quad \left[1 - \frac{2k^2a^2\, m_1m_2}{(m_1 + m_2)^2}\right] = \left[1 - \frac{4k^2a^2m_1m_2}{(m_1 + m_2)^2}\right]^{1/2}$$

Therefore, equation (3.20) becomes

$$(\omega_-^2) = f\left(\frac{m_1 + m_2}{m_1 m_2}\right)\left(1 - 1 + \frac{2k^2 a^2 m_1 m_2}{(m_1 + m_2)^2}\right)$$

$$= \frac{2fk^2 a^2}{m_1 + m_2}$$

$$(\omega_-) = ka\left[\frac{2f}{m_1 + m_2}\right]^{1/2}$$

$$= ka\left[\frac{2f}{m_1}\right]^{1/2} \quad \text{since } m_1 > m_2$$

$$(\omega_-) = \left(\frac{2f}{m_1}\right)^{1/2} \quad \text{for } k = \frac{\pi}{2a}$$

This gives acoustical branch (low frequency).

The range of 1st Brillouin zone is $-\dfrac{\pi}{2a} \le k \le \dfrac{\pi}{2a}$, where 2a is the distance between two atoms of the same nature. The complete curve of optical and acoustical branches corresponding to diatomic lattice are as shown in Fig. 3.5.

Some Important Features :

(i)   For optical branch, when $k \to 0$ then from equation (3.16), we get

$$(m_1 \omega^2 - 2f) A + 2Bf \cos ka = 0$$

But $\cos ka = 1$

$$\therefore \quad (m_1 \omega^2 - 2f) A + 2Bf = 0$$

$$\therefore \qquad \frac{\omega^2}{2f} = \frac{A - B}{m_1 A} \qquad \qquad \text{... (3.22)}$$

Similarly, from equation (3.15), we can write,

$$\frac{\omega^2}{2f} = \frac{B - A}{m_2 B} \qquad \qquad \text{... (3.23)}$$

From equations (3.22) and (3.23), we get

$$\frac{A - B}{m_1 A} = \frac{B - A}{m_2 B}$$

$$\therefore \qquad \frac{A}{B} = -\frac{m_2}{m_1}$$

This shows that two masses m_1 and m_2 move in opposite direction as shown in Fig. 3.6 (a) and their amplitudes being inversely proportional to the masses.

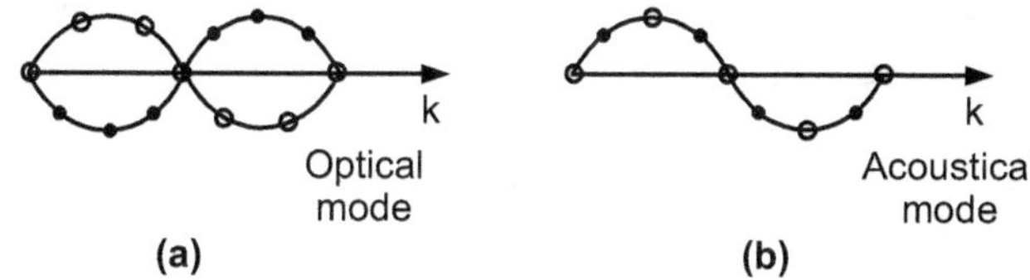

Fig. 3.6

(ii) For acoustical branch when $k \to 0$ then the second term in equation (3.15) becomes significant and $\cos ka \Rightarrow 1 - \dfrac{k^2 a^2}{2}$.

$$(m_2 \omega^2 - 2f)\, B + 2f\, A\left[1 - \frac{k^2 a^2}{2}\right] = 0$$

$$m_2 \omega^2 B + 2fA\left[1 - \frac{k^2 a^2}{2}\right] - 2Bf = 0 \qquad \ldots (3.24)$$

Again from equation (3.16), we can write

$$(m_1 \omega^2 - 2f)\, A + 2fB\left[1 - \frac{k^2 a^2}{2}\right] = 0 \qquad \ldots (3.25)$$

$$m_1 \omega^2 A + 2fB\left[1 - \frac{k^2 a^2}{2}\right] - 2fA = 0$$

Adding equations (3.24) and (3.25), we get

$$\omega^2\,[m_1 A + m_2 B] = k^2 a^2 \cdot f\,(A + B)$$

For acoustical branch, let us substitute $(\omega) = ka\left[\dfrac{2f}{m_1 + m_2}\right]^{1/2}$

$$\therefore \quad k^2 a^2\left[\frac{2f}{m_1 + m_2}\right][m_1 A + m_2 B] = k^2 a^2 f\,(A + B)$$

$$\therefore \qquad \qquad \frac{A}{B} = 1$$

This shows that the two masses vibrate in the same direction [Fig. 3.6 (b)] with same amplitude.

(iii) It is observed that when atoms of the lattice are of equal masses then the frequency range $0 < \omega^2 < \dfrac{4f}{m}$ is same for monoatomic as well as diatomic lattice. However, in monoatomic lattice, the whole range corresponds to acoustical branch.

In case of diatomic lattice, the range $a < \omega^2 < \dfrac{2f}{m}$ corresponds to acoustical branch while the range $\dfrac{2f}{m} < \omega^2 < \dfrac{4f}{m}$ corresponds to optical branch and there is no forbidden gap between two.

(iv) When m_2 decreases the optical branch shifts upwards but the acoustical branch is not affected and we return to the monoatomic lattice with lattice constant 2a.

(v) As m_1 increases, the acoustical branch yields downward and optical branch tends to flatten. When $m_1 \to 0$ the acoustical branch yields downward completely and optical branch flattened completely.

(vi) The graph of (A/B) amplitude ratio versus k i.e. wave vector is as shown in Fig. 3.7.

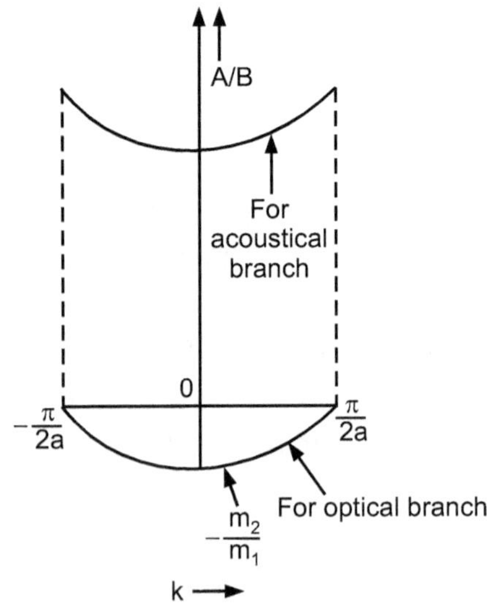

Fig. 3.7

1. At $k = \pm \dfrac{\pi}{2a}$ the ratio $\dfrac{A}{B} = 0$ for optical branch and ∞ for acoustical branch.

2. It has been observed that in acoustical branch the light particles of masses m_2 are all at rest ($B = 0$) and in optical branch the heavy particles of mass m_1 are at rest ($A = 0$).

3. If we excite the vibration of optical branch by a force that makes the two neighbouring atoms move in opposite direction. Because of this the branch is called as optical branch.

4. If we excite the vibration of acoustical branch by a force that makes the atoms move in the same direction. Because of this, the branch is called as acoustical branch.

(vii) **Forbidden energy gap :** From Fig. 3.5, it is seen that there is band of frequencies shown by shaded region between optical and acoustical branch that cannot propagate. This is called as forbidden band whose width depends upon mass difference. If $m_1 = m_2$, the forbidden band disappears at $k = \dfrac{2\pi}{a}$. This is characteristic feature of elastic wave in diatomic lattice. The frequency gap between two branches becomes wider as the mass ratio $\left(\dfrac{m_1}{m_2}\right)$ becomes larger. The optical band is found to become narrower with increasing ratio $\left(\dfrac{m_1}{m_2}\right)$. The wave vector (k) is found to be complex for frequencies in the forbidden band showing that the wave is damped in the space by the lattice.

3.5 EXPERIMENTAL DETERMINATION OF DISPERSION RELATIONS : [FOR LATTICE VIBRATIONS BY INELASTIC NEUTRON SCATTERING]

Dispersion relations for lattice vibration can be verified by impressing sound wave on the surface of crystal. However, it is not possible in practice, because the requirement of very high frequency, which have not been made so far. For verifying the dispersion relation, k should be equal to $\dfrac{\pi}{a}$ (Brillouin zone boundary) or $k = \dfrac{2\pi}{\lambda} = \dfrac{\pi}{a}$ or $\lambda = 2a$. As the velocity of sound is almost equal to 3.5×10^5 cm and the frequency 5×10^{12} C/s.

Generating such high frequency with strong intensity is not possible, hence direction of verification of dispersion relation does not seem possible.

Dispersion relations for lattice vibration are obtained by neutron diffraction. In neutron diffraction experiment, a burst of monoenergetic neutrons hits the crystal and the scattering occurs at Bragg's angle (elastic scattering) and scattering also occurs from interaction with lattice vibration (inelastic scattering). Neutron diffraction apparatus is as shown in Fig. 3.8.

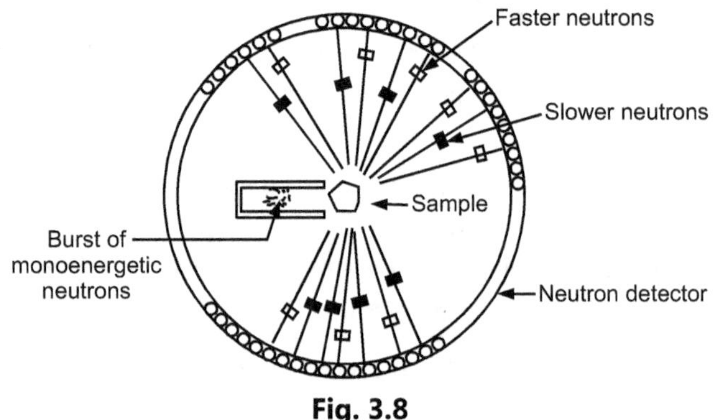

Fig. 3.8

In this apparatus a number of detectors are placed at the circumference of a circle in the plane containing the neutron source and the crystal. The purpose of this detector is to measure the energy and direction of the diffracted neutrons at various angles. A typical diffraction pattern obtained in the neutron diffraction experiment is as shown in Fig. 3.9. Its interpretation is based on theoretical considerations.

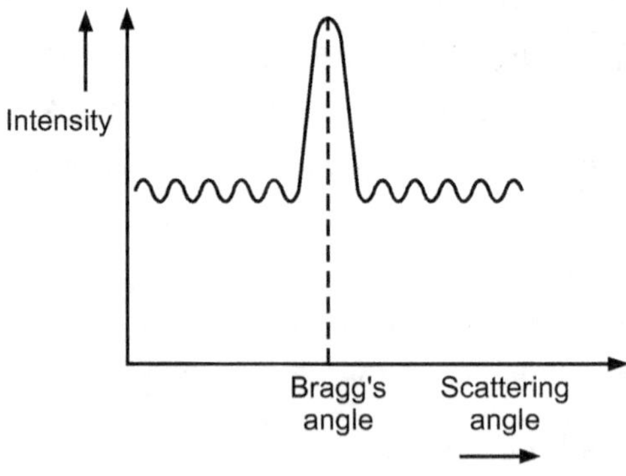

Fig. 3.9

(i)   In X-ray scattering from crystals the diffracted beam appears at Bragg's angle. The scattering is effective and the selection rule is

$$K' = K + G \qquad \ldots (3.26)$$

In scattering of neutrons from the crystals, the scattered beams at Bragg's angle are because of elastic scattering and follows the selection rule. In equation (3.26), K' and K are wave vectors of the scattered and incident photons respectively and G is the reciprocal lattice vector and is called as crystal momentum.

(ii)   The scattering of neutron is inelastic and accompanied by absorption or excitation of phonon, then the selection rule becomes

$$K' = K \pm q + G \qquad \ldots (3.27)$$

where $(+)$ means creation of phonons and $(-)$ means absorption of phonons.

(iii)   If the dispersion relations are determined by inelastic scattering of neutrons with emission or absorption of phonons then the rule of energy conservation must also be satisfied as

$$\frac{\hbar^2 k^2}{2m} = \frac{\hbar^2 k'^2}{2m} \pm \hbar k \qquad \ldots (3.28)$$

The strongest peak in the neutron diffraction is because of elastic diffraction and should occur at Bragg's angle and for this peak, the selection rule will be valid by equation (3.26). Both K and K' can be obtained from most intense peak [Refer Fig. 3.9].

Knowing the structure of the crystal w.r.t. to the incident neutrons, the wave vectors K and K' of the neutrons corresponding to the intense peak can be evaluated and hence G can be evaluated.

Now from small peak (Fig. 3.9) which are because of inelastic scattering of neutron by lattice vibration and with equation (3.27), K and K' can be evaluated and G is already known, q can be determined. Thus from equation (3.27), using the momentum conservation rule, the wave vector of the created or absorbed phonon can be calculated but its frequency ω is calculated by the rule of conservation of energy. The equation (3.28) can also be written as,

$$E = E' \pm \hbar\omega$$

where, $E = \dfrac{\hbar^2 k^2}{2m}$ and $E' = \dfrac{\hbar^2 k'^2}{2m}$, E is the energy of the monochromatic beam and E' is measured by the time of flight measurement. Thus with q (wave vector of the phonon), from equation (3.27) and from equation (3.28) wave vector-frequency ($\omega - k$) relation from observations in various detectors can be obtained and plotted in the Brillouin zone. Every detector distributed around the circle gives different set of ω_i and k_i to plot the dispersion curve. Both the transverse and longitudinal branches appear in the acoustic or optical regions.

SOLVED PROBLEMS

Problem 3.1 : Calculate the frequency of acoustic waves of wavelength 10^{-7} cm on a linear lattice containing two identical atoms per primitive cell of interatomic spacing 2.5 A° and light waves of same wavelength, given that $v_o = 10^5$ cm/sec.

Solution : For acoustic waves in diatomic lattice the frequency varies from $\omega = 0$ for $k = 0$ to $\omega = \sqrt{\dfrac{2f}{m}}$ for $k = \dfrac{2\pi}{a}$.

In case of diatomic lattice, we have

$$v_o = \sqrt{\dfrac{2f}{m}} \cdot a$$

where f is constant of force and m is mass of the particle and a is the interatomic distance.

$$\frac{v_o}{a} = \sqrt{\dfrac{2f}{m}}$$

$$\omega = \frac{v_o}{a} = \frac{10^5}{2.5 \times 10^{-8}} = 4 \times 10^{12} \text{ rad/sec}$$

For light waves of wavelength 10^{-7} cm, the velocity $c = 3 \times 10^{10}$ cm/s.

$$\therefore \quad \omega = 2\pi\upsilon = 2\pi\frac{c}{\lambda} = \frac{2\pi \times 3 \times 10^{10}}{10^{-7}}$$

$$= 6\pi \times 10^{17} \text{ rad/sec}$$

Problem 3.2 : The unit cell site of NaCl is 5.6 A° and Young's modulus in the (100) direction is 5×10^{10} Nm^{-2}. Estimate the wavelength at which electromagnetic radiation is strongly reflected by NaCl crystal.

[Atomic weight of Na 23, Cl = 37]

Solution : The frequency of radiation strongly reflected by ionic crystal is given by

$$\omega^2 = 2f\left[\frac{1}{m_1} + \frac{1}{m_2}\right]$$

where m_1 and m_2 are masses of atoms and $f = \dfrac{a}{2E}$.

where a is the interatomic distance and E is Young's modulus.

$$\omega^2 = a \cdot E\left[\frac{1}{m_1} + \frac{1}{m_2}\right]$$

$$= 5.6 \times 10^{-10} \times 5 \times 10^{10}\left[\frac{1}{23} + \frac{1}{37}\right] \times \frac{1}{1.67 \times 10^{-27}}$$

$$= \frac{5.6 \times 5 \times 60 \times 10^{27}}{23 \times 37 \times 1.67}$$

$$= 2.17 \times 10^{13} \text{ rad/sec}$$

The wavelength at which the electromagnetic radiation is strongly reflected is given by

$$\lambda = \frac{c}{\upsilon} = \frac{2\pi c}{\omega}$$

$$= \frac{2\pi \times 3 \times 10^8}{2.17 \times 10^{13}}$$

$$= 8.648 \times 10^{-5} \text{ m}$$

$$= \mathbf{86.48\ \mu}$$

EXERCISES

(A) Multiple Choice Type Questions :

1. If two masses m_1 and m_2 are equal then the frequency range in both monoatomic and diatomic lattice is ......

    (a) same   (b) different

    (c) less than other   (d) greater than other

2. The phase velocity (v_p) and group velocity (v_g) of the elastic wave can be different from each other in a ......

   (a) air medium

   (b) dispersive or crystal medium

   (c) colourful medium

   (d) continuous and homogeneous medium

3. The optical branch is active in ....

   (a) ultraviolet absorption   **(b) infra red absorption**

   (c) far-infra red absorption   (d) none of these

4. When elastic wave travels through the monoatomic crystal, it will act as ......

   (a) low pass filter   (b) high pass filter

   (c) band stop filter   (d) band pass filter

5. When elastic wave travels in a diatomic crystal, then in optical case all the atoms of the crystal move in ......

   (a) same direction   **(b) opposite direction**

   (c) transverse direction   (d) longitudinal direction

6. When elastic wave travels in a diatomic lattice, the width of the forbidden band depends upon ......

   (a) addition of masses   (b) multiplication of masses

   (c) difference of masses   (d) division of masses

7. The elastic waves are in a diatomic lattice. If two masses are the same then forbidden band disappears at ......

   (a) $k = +\dfrac{\pi}{2a}$ (b) $k = -\dfrac{\pi}{2a}$

   (c) $k = \sqrt{\dfrac{\pi}{2a}}$ (d) $k = \pm\dfrac{\pi}{2a}$

8. When elastic wave travels in a diatomic crystal medium then acoustical atoms of the crystal move in ......

   (a) longitudinal direction

   (b) opposite direction

   (c) same direction

   (d) transverse direction

9. In case of steady wave, the boundaries of the first Brillouin zone behave like ......

   (a) nodal points

   (b) antinodal points

   (c) longitudinal points

   (d) transverse points

10. For acoustical branch ...... at $k = 0$, when A and B are amplitudes of vibration of particles.

    (a) $A \neq B$

    (b) $A = B$

    (c) $A = -B$

    (d) none of these

(B) Short Answer Type Questions :

1. Explain group velocity and phase velocity.

2. Explain optical and acoustical properties of the crystal.

3. Write a note on forbidden frequency band.

4. For optical branch, show that $m_1 A = m_2 B$ at $k = 0$, where A and B are amplitudes of vibration of particles.

5. For acoustical branch, show that $A = B$, at $k = 0$, where A and B are amplitudes of vibration of particles.

(C) Long Answer Type Questions :

1. Discuss the vibration of monoatomic linear lattice and hence define phase velocity and group velocity.

2. Discuss the vibration of diatomic linear lattice and show that vibrational spectrum consists of two branches.

3. For linear diatomic lattice, obtain an expression for two frequencies denoted by (ω_+) and (ω_-).

4. Explain how crystal behave like a low pass filter in one dimensional monoatomic lattice vibration.

5. Discuss vibration of monoatomic linear lattice and hence obtain expression for frequency.

6. Obtain dispersion relation in the vibration of diatomic linear lattice.

7. Discuss experimental determination of dispersion relation for lattice vibration by inelastic neutron scattering.

4

CHAPTER

FREE ELECTRON THEORY OF METALS AND BAND THEORY OF SOLIDS

SYLLABUS

Sommerfield's free electron model for electrical conductivity of metals, Fermi-Dirac distribution, Origin of energy bands - Valence band, Conduction band, Band gap energy, Distinction between metals, semiconductors and insulators, Hall effect – Hall voltage and Hall coefficient.

4.1 INTRODUCTION

We know that the crystals-metals are made up of regular arrangement of atoms. Due to the regular arrangement of atoms, it is possible to study a partial theoretical understanding of their properties by the application of wave mechanical principles. There are two ways to study the metals by wave mechanics. In one, the metal is regarded as made up of atoms in mutual interaction, in the other metal is regarded as composed of nuclei and electrons. The valence electrons are loosely bound to their individual atoms in metals. They move randomly in different directions and are called as free electrons. The binding forces on these electrons are very small and they behave like free electrons in metals. It is noticed that except free electrons, all electrons are tightly bound to the nuclei. The valence electrons of atoms became the conductors of electricity in metals and they are known as conduction of electrons.

The outstanding properties of metals are

(1) High electrical and thermal conductivities.

(2) In steady state, Ohm's law is obeyed.

(3) Resistivity above Debye's temperature is proportional to absolute temperature.

(4.1)

(4) At low temperature the resistivity of metals is proportional to 5th power of absolute temperature.

(5) For most of metals the resistivity is inversely proportional to pressure.

(6) Above Debye temperature the ratio of thermal and electrical conductivities are proportional to absolute temperature.

(7) At absolute zero, the resistivity tends towards zero (superconductivity).

4.2 SOMMERFIELD'S FREE ELECTRON MODEL FOR ELECTRICAL CONDUCTIVITY OF METALS

To overcome the difficulties in electron gas model of metals, Sommerfield in 1928 suggested the new model in which potential is taken as the constant inside the metal so that no force is acting on the electron. An electron in a metal finds itself in the field of all nuclei and all other electrons. The potential energy for such an electron may therefore be expected to be periodic. In the model employed by Sommerfield, however, it is assumed that the free electrons i.e. those giving rise to conductivity find themselves in a potential which is constant everywhere inside the metal. Since one does not observe electron emission from metals at room temperature, it seems evident that the potential energy of an electron at rest must be lower than that of an electron at rest outside the metal.

In Sommerfield's physical model of solid, the interior of the metal is represented by potential energy box. In this model, the free electrons are assumed to be valence electrons of the composing atoms. Thus the alkali metals are assumed to contain one free electron per atom. He investigated the behaviour of free electron gas firstly by discovering the possible energy states for an electron in the potential energy box and secondly by considering the distribution of large number of electrons in thermal equilibrium among the states.

For the sake of mathematical simplicity, consider first a free electron of mass (m) limited to remain within one dimensional crystal of length (L). Next assume that the potential energy every where, within the crystal (box) is constant and is equal to zero i.e. V = 0. At the two ends of the crystal the free electron is prevented from leaving the crystal by very high potential energy barrier (V_0) as shown in Fig. 4.1.

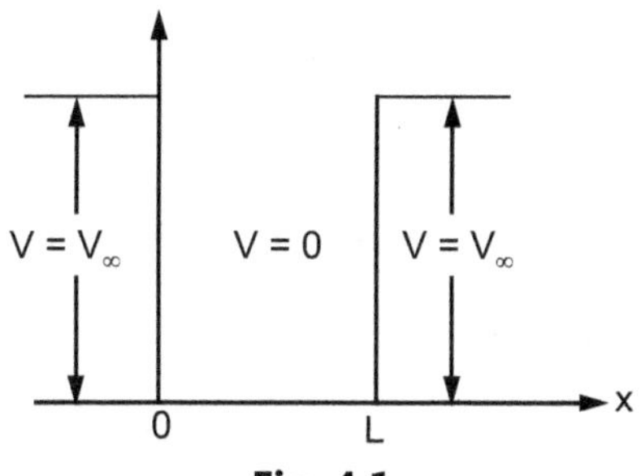

Fig. 4.1

Here the boundary conditions are

$$V(x) = 0 \text{ when } 0 < x < L$$

$$= \infty \text{ when } x \leq 0 \text{ and } x \geq L$$

Since the potential energy inside the crystal is zero, the Schrödinger's equation has the form

$$\frac{d^2\psi_n}{dx^2} + \frac{2mE_n}{\hbar^2}\psi = 0$$

where E_n is the energy in n^{th} state of an electron.

$$\frac{d^2\psi_n}{dx^2} + \alpha^2 \cdot \psi_n = 0 \qquad \qquad ... (4.1)$$

where, $$\alpha^2 = \frac{2mE_n}{\hbar^2}$$

This is the second order differential equation with variable x. The general solution of the equation (4.1) is,

$$\psi_n(x) = A \sin \alpha x + B \cos \alpha x \qquad \qquad ... (4.2)$$

where A and B are constants to be determined by boundary conditions.

$$\therefore \qquad \psi_n(x) = 0 = A \sin (\alpha \times 0) + B \cos (\alpha \times 0), \text{ at } x = 0$$

$$= B \qquad \qquad (\text{since } \cos 0 \neq 0)$$

Again at x = L, the equation (4.2) becomes

$$\psi_n(x) = 0 = A \sin \alpha L$$

i.e. $$A \sin \alpha L = 0 \qquad \qquad (\text{since } A \neq 0)$$

$$\therefore \qquad \sin \alpha L = 0$$

i.e. $\qquad \alpha L = \pi,\ 2\pi,\ 3\pi,\ ... = n\pi$ where $n = 1, 2, 3 ...$

$\therefore \qquad \alpha = \dfrac{n\pi}{L}$ ... (4.3)

$\therefore \qquad \psi_n(x) = A \sin \dfrac{n\pi}{L} \cdot x$

$$= \sqrt{\dfrac{2}{L}} \cdot \sin \dfrac{n\pi}{L} \cdot x \qquad ... (4.4)$$

The constant A is calculated by normalisation condition and found to be $\sqrt{\dfrac{2}{L}}$.

The equation (4.3) can be written as

$$\alpha^2 = \dfrac{n^2\pi^2}{L^2} \qquad ... (4.5)$$

From equations (4.1) and (4.5), we get

$$\dfrac{2mE_n}{\hbar^2} = \dfrac{n^2\pi^2}{L^2}$$

$\therefore \qquad E_n = \dfrac{n^2\pi^2}{2mL^2} \cdot \hbar^2 = \dfrac{\hbar^2}{2m}\left(\dfrac{n\pi}{L}\right)^2$

$$= \dfrac{n^2\pi^2}{2mL^2} \cdot \dfrac{h^2}{4\pi^2} \qquad \left(\text{since } \hbar = \dfrac{h}{2}\right)$$

$$= \dfrac{n^2 h^2}{8mL^2} \qquad ... (4.6)$$

From this equation it is seen that the energy can have only the discrete values corresponding to n = 1, 2, 3 ... etc. called as quantized energy values.

The parabolic variation of energy as a function of quantum number (n) is as shown in Fig. 4.2.

Although the energy variation is drawn as a continuous curve, it actually consists of discrete points, corresponding to the discrete values of E_n. Because adjacent energy values differ by less than 10^{-18} eV it has not possible to show actual breaks in this quasi-continuous energy distribution (Refer Fig. 4.2).

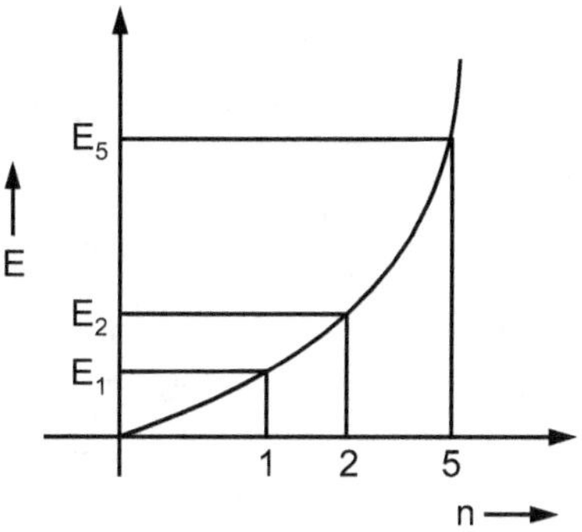

Fig. 4.2

In three dimensions, a crystal can be approximated by a cube of edge (L) inside which the potential energy is zero. It can be shown that the wave function inside such cube is given by [using equation (4.4)].

$$\psi_n(r) = \left(\frac{8}{L^3}\right)^{1/2} \cdot \sin\left(\frac{n_x\pi}{L}\right) x \cdot \sin\left(\frac{n_y\pi}{L}\right) y \sin\left(\frac{n_z\pi}{L}\right) z \qquad \ldots (4.7)$$

where n_x, n_y and n_z are three integers greater than zero.

The corresponding form of energy in three dimensions is

$$E_{n_x\,n_y\,n_z} = \frac{h^2}{8mL^3}\left[n_x^2 + n_y^2 + n_z^2\right] \qquad \ldots (4.8)$$

The form of equations (4.7) and (4.8) are identical except for the number of integers. The quantity $R^2 = n_x^2 + n_y^2 + n_z^2$ is called as momentum lattice vector drawn from origin to any lattice point in the momentum space.

4.3 MOMENTUM SPACE

The kinetic energy of an electron can be related to its momentum by $E = \dfrac{p^2}{2m}$ and the potential energy is assumed to be zero everywhere inside the metal. Consequently the total energy of the electron is equal to its kinetic energy.

$$\therefore \qquad E_{n_x\,n_y\,n_z} = \text{K.E. of free electron}$$

$$= \frac{1}{2} mv^2 = \frac{1}{2}\frac{m^2v^2}{m} = \frac{p^2}{2m}$$

Therefore equation (4.8) becomes

$$E_{n_x\,n_y\,n_z} = \frac{h^2}{8mL^2}\left[n_x^2 + n_y^2 + n_z^2\right] = \frac{P^2}{2m} \qquad \text{... (4.9)}$$

$$\therefore \qquad R^2 = \left(\frac{2L}{h}\right)^2 \cdot P^2 = n_x^2 + n_y^2 + n_z^2 \qquad \text{... (4.10)}$$

From this it follows that the momentum of an electron can be expressed in terms of three quantum numbers. From equations (4.9) and (4.10), it follows that as length R increases, so does the momentum and hence the energy of the state is represented by lattice point. In free electron model of a metal containing about 10^{23} electrons, lattice points are so closely spaced that it is not possible to show all of them in picture.

4.4 FERMI-DIRAC DISTRIBUTION

The free electron model of a metal has survived to the present time because it is fairly close approximation to the actual situation in metals particularly the monovalent elements such as alkali metals. Quantum mechanics requires that all valence or free electrons should be indistinguishable but that the state of each electron be specified by three quantum numbers n_x, n_y, n_z together with spin which can have either of two values $\pm\frac{1}{2}$. Moreover, the Pauli's exclusion principle does not permit more than one electron to have the same four quantum numbers. By using quantum statistics, Fermi-Dirac derived the formula to state the probability F(E) that a particular quantum state is having energy (E) occupied by the free electron and is given by

$$F(E) = \frac{1}{e^{\frac{E - E_F}{kT}} + 1} \qquad \text{... (4.11)}$$

This is called as F.D. distribution formula.

Where E_F is Fermi-energy of the metal, k is Boltzmann constant and T is the temperature of metal.

The distribution of free electrons in the metal can be studied by F.D. distribution as :

(i)   At T = 0 and E << E_F, the exponential term in equation (4.11) is very small so that F(E) is essentially equal to unity.

i.e. for E << E_F,   F(E) = 1

This means that all the states having energy smaller than Fermi-energy are completely occupied.

(ii) At T = 0 but E >> E_F, the exponential term is very large and F(E) rapidly tends to zero as energy increases.

i.e. for E >> E_F,   F(E) = 0

This means that all the states having energy greater than Fermi-energy are virtually empty.

(iii) If T > 0 i.e. T ≠ 0 and E = E_F, then F(E) = $\frac{1}{2}$.

This means that the states are half filled (50%) and half empty (50%).

The actual distribution of free electrons is as shown in Fig. 4.3 and it is evident that the higher energy states are virtually empty.

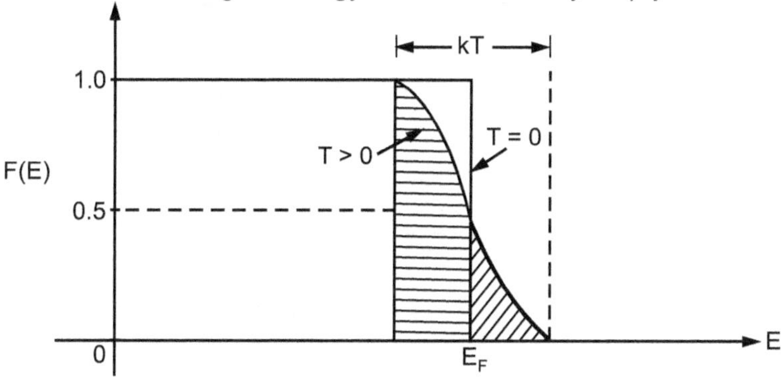

Fig. 4.3

The important consequence of applying quantum statistics to free electrons in a metal is that it is now possible to predict correctly how many electrons can gain energy from an external source. The amount of energy that an electron can gain from either a thermal source or magnetic field is of the order of kT. [At room temperature, 1 kT = 0.03 eV]. Only the electrons whose energy is already very close to the Fermi energy level can actually gain more energy because electrons occupying lower energy levels would have to undergo transitions to already occupied states. The fraction of electrons, that can undergo such transitions can be deduced by considering the shaded region in Fig. 4.3 since the height of the curve is unity. The ratio of the shaded area to the total area under the curve is

$$\frac{k.T.}{E} = \frac{0.03}{3.00} = \frac{1}{100}$$

Selecting 3.00 eV as a representative value of the Fermi energy in a metal.

4.5 THE FERMI ENERGY (E_F)

We know that at absolute temperature, all the states having energy less than E_F are completely filled ($E_F = 1$) and those having energy greater than E_F are completely empty i.e. $E_F = 0$. Thus the maximum energy that an electron can have at absolute zero is the Fermi energy E_F. The actual value of this energy can be determined directly by making the use of the lattice point in momentum space. According to this concept, the lattice point represents the allowed states at absolute zero are occupied upto some maximum value of (R). Since each state can be occupied by two electrons having opposite spins; so there are N/2 occupied states in crystal containing N free electrons. The lattice point representing these states lie in the positive octant of the lattice because n_x, n_y, n_z can have positive values and are enclosed by a sphere of radius R_{max} given by the equations (4.8) and (4.10) as

$$E_F = \frac{h^2}{8mL^2} R_{max}^2$$

When
$$R = R_{max} \text{ and } E_{n_x n_y n_z} = E_F$$

$\therefore$
$$R_{max} = \left[\frac{8mL^2 E_F}{h^2} \right]^{1/2}$$

Therefore, the number of occupied states in the metal is at absolute zero and is given by

$$\frac{N}{2} = \frac{1}{8} \times \text{Volume of the sphere of radius } R_{max}$$

$$= \frac{1}{8} \left[\frac{4}{3} \pi R_{max}^3 \right]$$

$$= \frac{1}{8} \left[\frac{4}{3} \pi \left(\frac{8mL^2 E_F}{h^2} \right)^{1/2} \right]$$

$$= \frac{\pi}{6} \left(\frac{8mL^2 E_F}{h^2} \right)^{1/2}$$

$$\frac{N}{2} = \frac{\pi L^3}{6} \left(\frac{8mE_F}{h^2} \right)^{3/2}$$

$\therefore$
$$\frac{8m \cdot E_F}{h^2} = \left(\frac{3N}{\pi L^3} \right)^{2/3}$$

$$\therefore \qquad E_F = \frac{h^2}{8m}\left(\frac{3N}{\pi L^3}\right)^{2/3}$$

$$= \frac{\hbar^2}{8m}\left(\frac{3n}{\pi}\right)^{2/3}$$

$$E_F = E_{F_0} = \frac{\hbar^2}{2m}(3\pi^2 n)^{2/3} \qquad \qquad \ldots (4.12)$$

where E_F is independent of temperature of the metal.

Here, $\qquad n = \dfrac{N}{L^3} = \dfrac{\text{Number of free electrons in the metal}}{\text{Volume of the metal}}$

$$= \text{Density of the metal}$$

From equation (4.12) it is seen that E_F is a function of $\dfrac{N}{L^3}$ i.e. number of free electrons per unit volume i.e. density of metals, since all other factors on right hand side of equation (4.12) are constant. Consequently the Fermi energy does not change when two identical metals are joined together i.e. value of Fermi energy is independent of the size of the metal. The calculation of Fermi energy in a metal is strictly correct only at absolute zero of temperature.

Actually the Fermi level is not constant when electron energy distribution in metal crystals are conducted but changes slightly with increasing temperature. Let $E_{F(0)}$ is the value of Fermi energy at absolute zero, then its level at any other temperature is given by,

$$E_F = E_{F(0)} = \left[1 - \frac{\pi^2}{12}\left(\frac{kT}{E_{F(0)}}\right)^{1/2}\right]$$

At room temperature the change in Fermi level is only about 0.2%, so that its effect is not important. At elevated temperature, however, the decrease in E_F is significant.

4.6 ORIGIN OF ENERGY BANDS – VALENCE BAND, CONDUCTION BAND, BAND GAP ENERGY

A solid contains an enormous number of atoms together. Each atom when isolated has a discrete set of energy levels 1s, 2s, 2p ... etc. If we imagine all N atoms of the solid to be isolated from one another, they would have completely coincide schemes of their energy levels. The energies of electrons within any one isolated atom obeyed following conditions.

(i)   There are specific electronic energy levels around each atom [Refer Fig. 4.4 (a)]. Electrons cannot occupy spaces between these levels.

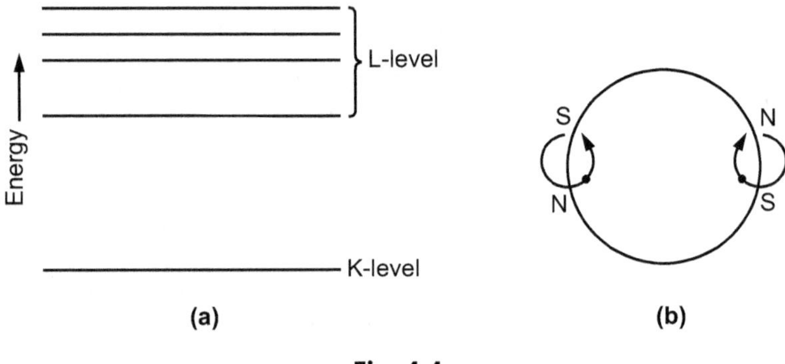

(a)　　　　　　　　　　　　(b)

Fig. 4.4

(ii) Electrons fill the lowest energy level first. A specific quantity of energy called a quantum of energy, must be supplied to move an electron to the next higher level.

(iii) According to Pauli's exclusion principle, no two electrons can occupy the same quantum states. Not more than two electrons may occupy any one energy level. Two electrons may occupy the same energy level because they have opposite magnetic directions [Refer Fig. 4.4 (b)].

Let us see what happens to the energy levels of isolated atoms as they are brought closer together to form a solid. If the atoms are brought in close proximity the valence electron of the adjacent atoms interact. Hence valence electrons constitute a single system to entire crystal. Therefore N electrons will occupy different energy levels. This is brought by electric forces exerted on each electron by all N nuclei. As a result of these forces, each atomic energy level is split up into larger number of energy levels. A set up of such energy level is called as energy band.

Consider an example of sodium which contains eleven electrons each occupying specific energy level as shown in Fig. 4.5. In Fig. 4.5, r_o represents the spacing between atoms in solid sodium. When the atoms are the part of solid, they interact with each other and the electrons have slightly different energies.

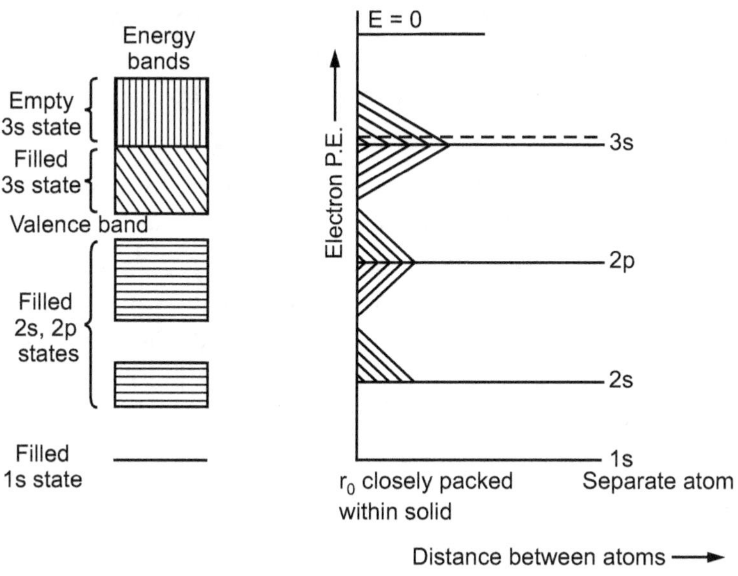

Fig. 4.5

As the spacing between the atoms is of the order of interatomic distance, the resultant energy level will be different than that of isolated atom. In Fig. 4.5, it is seen that quantum state 3s is located above 2p state. This means that valence electron is no longer bound to its parent nucleus but it is free to move inside the crystal which acts as free electron. But they cannot leave the crystal because their energy is not enough to climb the potential barrier at the surface of crystal.

At the surface extra atom gives energy distribution which rises upto E = 0 and this offers barriers to the free electrons. The free electrons can move randomly inside the crystal but the net charge at the location is equal to zero. The electron in the quantum states 2p, 2s and 1s are not free to move inside the crystal. Such electrons are called as bounded electrons. Fig. 4.5 shows the splitting of energy levels (3s) corresponding to five sodium nuclei. As there is one valence electron per nucleus, this gives rise to five free electrons. Due to wave mechanical coupling, the splitting would lead to five allowed energy levels and ten quantum states. Similarly the electron occupying 2p energy level will also experience wave mechanical coupling which is much weaker as compared to valence electrons, hence splitting at 2p would be much less. The electrons at 2s and 1s are well shielded and hence these energy levels exhibit no splitting.

The coupling depends only upon interatomic distance which is constant for crystal. Therefore, difference in energies would remain constant for sodium, whatever be the size of the crystal. If the crystal size is 100 c.c. we would have 100×10^{22} energy levels in the same energy difference. As there are large number of energy levels in a narrow width, we cannot speak about individual energy levels but must talk in terms of two energy bands. The number of energy levels in this band depends upon the size of the crystal, so that density of energy level is constant. This means that the number of quantum states in a system is conserved preserving the Pauli's exclusion principle.

Valence band :

The energy occupied by the valence electron is called as valence band. It is lower band. This band may be either completely filled or partially filled with electrons but can never be empty. It is completely filled in case of insulators.

Conduction band :

The electrons which have left the valence band are called as conduction band. They practically leave the atom or only weakly bound the nucleus. The band occupied by these electrons is called as conduction band. This band lies above the valence band. It may be either empty or partially filled with conduction electrons. In conduction band the electrons move freely and conduct electric current through the solid. It is completely empty in case of insulator.

Forbidden energy gap :

The V.B. and C.B. are separated by energy gap known as forbidden energy gap. The width of forbidden energy gap (E_g) is the characteristic of the solid. Forbidden energy gap is nothing but the energy required to lift an electron from V.B. and put it into C.B. For example, E_g for Ge is 0.72 eV, for Si 1.12 eV.

4.7 DISTINCTION BETWEEN METALS, SEMICONDUCTORS AND INSULATORS

The proper distinction between metal, insulator and semiconductor can be made with the help of energy between valence band and conduction band. For this let us consider a particular energy band filled with electrons upto certain value k as shown in Fig. 4.6.

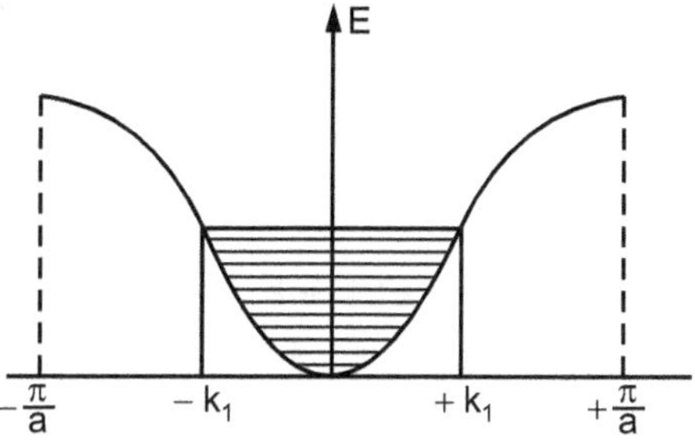

Fig. 4.6

The effective number of free electrons under the influence of external electric field is

$$N_{eff} = \Sigma F_k$$

where, F_k measures for the extent to which an electron in the state (k) is free to take part in electrical conduction and summation extends over occupied state in the band.

The number of states in the range dk for one dimensional lattice of length (L) is

$$d_n = \frac{L}{2\pi} \cdot dk$$

$$N_{eff} = \frac{L}{2\pi} \int\limits_{-k_1}^{+k_1} F_k \, dk \qquad \qquad \dots (4.13)$$

As there are two electrons occupying each of these states in the shaded region then the terms on right hand side of equation (4.13) are to be multiplied by 2, hence

$$N_{eff} = \frac{2L}{2\pi} \int_{-k_1}^{+k_1} F_k \, dk = \frac{L}{\pi} \int_{-k_1}^{+k_1} \frac{m}{\hbar^2} \frac{d^2E}{dk^2} \, dk$$

$$= \frac{mL}{\pi\hbar^2} \int_{-k_1}^{+k_1} \frac{d^2E}{dk^2} \, dk$$

$$= \frac{2mL}{\pi\hbar^2} \int_{0}^{k_1} \left(\frac{d^2E}{dk^2}\right) dk$$

$$N_{eff} = \frac{2mL}{\pi\hbar^2} \left[\frac{dE}{dk}\right]_{k=k_1}$$

The following conclusions can be drawn.

(i) At the top and bottom of the band i.e. at $k = 0$ and $\frac{\pi}{a}$ for E-k curve, $\frac{dE}{dk} = 0$ and thus effective number of electrons in completely filled band will be zero i.e. $N_{eff} = 0$.

(ii) Since $\frac{dE}{dk}$ becomes maximum at the point of inflection, the effective number becomes maximum for a band filled to the point of inflection.

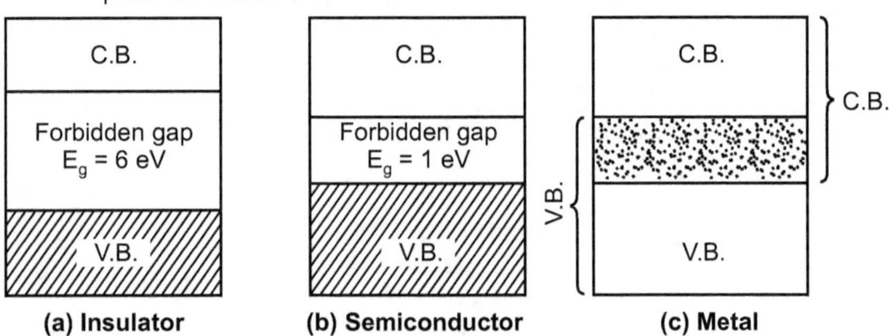

(a) Insulator   (b) Semiconductor   (c) Metal

Fig. 4.7

According to band theory of solids, the electrons are distributed in allowed energy bands. The outermost energy band having electrons is called as valence band while the next allowed empty band is called as conduction band. The electrons present in the conduction band only contribute to the electrical conductivity.

Insulator : The effective number of free electrons in the band will be zero i.e. $N_{eff} = 0$, when $\dfrac{dE}{dk} = 0$, this means that in the band there are no free electrons and all electrons are bound. Such a band is called as completely filled band and the other is completely empty. An insulator has wide forbidden energy gap. The valence band is completely filled with electrons and conduction band is completely empty. Obviously upper band cannot contribute to the electrical conductivity, because no electrons can cross the forbidden gap. This is an insulator. Refer Fig. 4.7 (a).

The examples of insulators are wood, glass, rubber etc.

Semiconductor : If the forbidden gap between filled (valence) band and empty (conduction) band is of the order of 1 eV as shown in Fig. 4.7 (b). At room temperature, few electrons can be promoted from valence band to conduction band across the forbidden energy gap. The electrons promoted to the conduction band can conduct electricity. When temperature is increased more electrons are excited into conduction band, hence conductivity of the semiconductor increases with increase of temperature or resistivity goes on decreasing with increase of temperature. Therefore semiconductor have negative temperature coefficient of resistance. The semiconductors are poorer conductors than metals but better than insulators. The examples are Ge, Si ... etc.

Metals (Good conductors) :

In metals the valence band and conductor band overlap because there is no forbidden energy gap. Refer Fig. 4.7 (c). On application of electric field the electrons may acquire additional energy and to move in higher energy state. Therefore, metals are excellent conductors. Electrons in the bands are distributed in accordance with Pauli's exclusion principle. At absolute zero all electrons fill up the lowest level and the highest filled level is called as Fermi level. All intrinsic semiconductors are insulators at T = 0, while all insulators may be treated as semiconductors at T > 0. The examples are Al, Fe, Cu or any metals.

4.8 HALL EFFECT

This effect was observed by H. Hall in 1879. The Hall effect is observed when a magnetic field is applied at right angles to a conductor carrying a current as shown in Fig. 4.8.

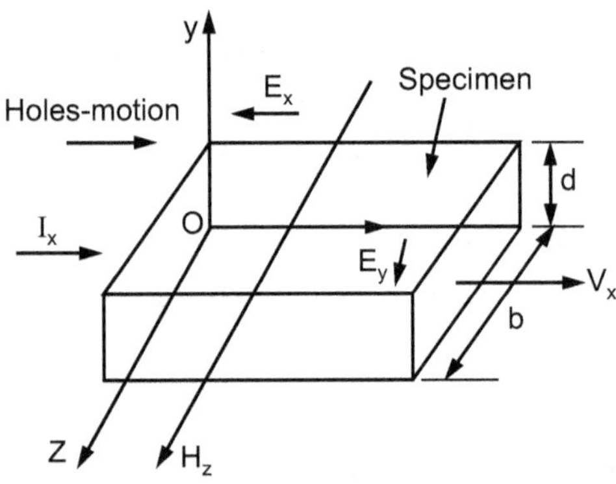

Fig. 4.8

The magnetic field gives an electric field in a direction mutually orthogonal to the direction of current and magnetic field. The electric field (E) produces a current (I) causing a force of magnitude eE to act on a electron. In the presence of magnetic field, a magnetic force proportional to the magnetic field strength (H) and average velocity (V) also act on the electrons. This force is at right angles to the directions of (H) and (V) and therefore each electron is deflected towards one side of the conductor. When electron reaches to the surface of the conductor, an electrical charge is produced. Under the equilibrium condition, the electron can again move freely down the conductor.

To study the Hall effect, let us consider a slab of material subjected to an external field (E_x) along x-axis and magnetic field (H_z) along z-axis as shown in Fig. 4.8. As a result of the applied electric field, a current density (I_x) will flow in a direction of x-axis. For the moment let us suppose that the current is carried by electrons of charge (−e). Under the influence of magnetic field, the electron will be subjected to a Lorentz force such that the lower surface collects a negative charge and upper surface collects positive charge. Finally when equilibrium state is reached the current along y-direction vanishes and field (E_y) is set up. If the charge carriers were positive the upper surface would become negative and lower surface is positive i.e. E_y would be reversed. Thus measure of Hall voltage gives the information about the sign of the charge carrier.

The identification of n-type and p-type semiconductor is done by knowing the polarity of the Hall voltage. If the polarity of Hall voltage is positive on the top surface of the specimen then the current I_z is due to the motion of electron and the specimen must be p-type. If the polarity of the Hall voltage on the top surface is negative the current I_x is due to holes and hence the specimen is p-type.

Consider the current I_x is due to the flow of electrons then electric force on electron having charge $-e$ is $-eE$.

The force due to magnetic field $H = -\dfrac{e}{c}(V \times H)$, where V is the average drift velocity of electrons and c is the velocity of light.

The total force on electron is

$$F = -e\left[E \times \frac{1}{c}V \times H\right]$$

In present case, $F_y = -e\left[E_y - \frac{1}{c}V_x\,H_z\right]$ ... (4.14)

4.9 HALL VOLTAGE AND HALL COEFFICIENT

In steady state the force due to accumulation of electrons becomes equal to the magnetic force and so the flow of electrons stop i.e. $F_y = 0$.

Therefore equation (4.14) becomes

$$0 = -e\left[E_y - \frac{1}{c}V_x\,H_z\right]$$

$\therefore$ $E_y = \dfrac{1}{c}V_x\,H_z$ ... (4.15)

This is called as Hall voltage and V_x is average drift velocity.

Further, the current density I_x may be expressed in terms of number of electrons per unit volume of the specimen as

$$I_x = -neV_x$$

$$\therefore \qquad V_x = -\frac{I_x}{ne}$$

Substitute the value of V_x in equation (4.15).

$$\therefore \qquad E_y = \frac{1}{c}\left(-\frac{I_x}{ne}\right)H_z$$

$$\therefore \qquad E_y = -\frac{1}{c}\frac{I_x}{ne}H_z$$

$$\therefore \qquad \frac{E_y}{I_x H_z} = -\frac{1}{nec} = R_H = \frac{1}{\rho c} \qquad\qquad \dots (4.16)$$

where R_H is known as Hall coefficient and $\rho = ne$ = charge density in the specimen.

$$\therefore \qquad R_H = -\frac{1}{\rho c} \text{ in e.s.u.}$$

$$= -\frac{1}{\rho} \text{ in e.m.u.}$$

The Hall coefficient is constant for a given material. It has unit of volt. cm. amp^{-1}. gauss^{-1}.

It is noted that the sign of the Hall coefficient is the same as the sign of the carrier. Thus it is negative, if the conduction is by electrons. If the conduction is based on holes, the sign of charge carriers has been positive and corresponding Hall coefficient would have been positive.

4.10 MOBILITY (μ), CONDUCTIVITY (σ) AND HALL ANGLE (ϕ)

When current carrying particles acquire a velocity per unit electric field, the velocity is known as mobility (μ).

$$\therefore \qquad \mu = \frac{V_x}{E_x}$$

$$\therefore \qquad V_x = \mu E_x$$

Substituting the value of V_x in equation (4.15), we get

$$E_y = \frac{1}{c}\mu E_x \cdot H_z \qquad\qquad \dots (4.16)$$

From equations (4.15) and (4.16), we get

$$R_H \cdot I_x \cdot H_z = \frac{1}{c}\mu E_x \cdot H_z$$

$$\therefore \qquad \mu = \frac{R_H \cdot I_x}{E_x} \cdot c$$

But

$$\sigma = \frac{I_x}{E_x} = \text{Electrical conductivity}$$

$$\therefore \qquad \mu = R_H \cdot \sigma \cdot c \text{ in e.s.u.}$$

$$= R_H \cdot \sigma \text{ in e.m.u.}$$

Again, we have,

$$\mu = R_H \cdot \sigma = \frac{E_y}{I_x H_x}\sigma$$

$$= \frac{E_y}{\sigma E_x \cdot H_x}\sigma$$

$$\therefore \qquad \mu = \frac{E_y}{E_x}\cdot\frac{1}{H_z} = \phi \cdot \frac{1}{H_z}$$

where,

$$\phi = \frac{E_y}{E_x} = \text{Hall angle}$$

$$\therefore \qquad \phi = \mu \cdot H_z$$

Again, we have,

$$\mu = R_H\sigma = -\frac{1}{\rho}\sigma$$

$$\therefore \qquad \sigma = -\rho \cdot \mu$$

Thus knowing conductivity (σ) and (R_H), the mobility (μ) of the current carriers can be calculated.

4.11 IMPORTANCE OF HALL EFFECT

(1) The sign of the current carrying charge is determined.

(2) The mobility is measured directly.

(3) The number of charge carriers per unit volume can be calculated from the Hall coefficient (R_H).

(4) It can be used to determine the electronic structure of the substance i.e. whether they are metals, semiconductors or insulators.

(5) The knowledge of Hall voltage developed enables us to measure high unknown magnetic field provided we know the Hall constant for the slab used for it.

EXERCISES

(A) Multiple Choice Type Questions :

1. The value of Fermi energy is independent of ...... of metal.

    (a) size　　　　　　　　(b)　mass

    (c)　temperature　　　　　(d)　length

2. At absolute zero of temperature, when $E << E_F$ then $F(E) = ...$

    (a)　1　　　　　　　　　　(b)　−1

    (c) 0　　　　　　　　　(d)　$\dfrac{1}{2}$

3. The sign of Hall coefficient R_H in Hall effect is ......

    (a)　same as the polarity of minority carriers in the semiconductor

    (b) same as the polarity of majority carriers in the semiconductor

    (c)　same as polarity of ions in crystals

    (d)　opposite as the polarity of majority carriers in the semiconductor

4. In intrinsic semiconductor the number of conduction electrons in C.B. and number of holes in V.B. are ......

    (a) exactly equal　　　　(b)　unequal

    (c)　large　　　　　　　　(d)　zero

5. Hall coefficient (R_H) is negative if the conduction is due to ......

    (a)　holes　　　　　　　　**(b)　electrons**

    (c)　both (a) and (b)　　　(d)　none of these

6. At absolute zero of temperature, when $E >> E_F$, then $F(E) = ...$

    (a) 1　　　　　　　　　(b)　−1

    (c)　$\dfrac{1}{2}$　　　　　　　(d)　$-\dfrac{1}{2}$

7.   All insulators may be treated as semiconductors at ......

   (a) T < 0   **(b) T > 0**

   (c) T = 0   (d) T ≥ 0

8.   Mobility of current carriers (electrons) is ......

   (a) reciprocal of conductivity

   (b) flow of electrons per unit electric field

   (c) average electron drift velocity per unit electric field

   (d) none of these

9.   The Hall coefficient (R_H) is positive if the conduction is due to ......

   (a) holes   (b) electrons

   (c) both (a) and (b)   (d) none of these

10. The width of forbidden band is wide in case of ......

   (a) metal   (b) semiconductor

   (c) insulator   (d) none of these

11. All intrinsic semiconductors are insulators at ......

   (a) T ≥ 0   (b) T < 0

   (c) T = 0   (d) T > 0

12. The Fermi energy of metal ......

   (a) increases with increase of temperature

   (b) decreases with decrease of temperature

   (c) decreases with increase of temperature

   (d) independent of change in temperature

13. In completely filled band, the number of free electrons are ......

   (a) infinity   **(b) zero**

   (c) small   (d) large

14. If the temperature of the metal is greater than zero and $E = E_F$ then Fermi energy F(E) is ......

   (a) $\dfrac{1}{2}$ (b) $-\dfrac{1}{2}$

   (c) 1   (d) -1

(B) Short Answer Type Questions :

1. Define Fermi energy. Derive its expression for absolute zero of temperature.

2. Explain F.D. distribution of electron in metal.

3. What do you mean by C.B., V.B. and forbidden energy gap ?

4. Discuss the theory of origin of energy bands in crystals.

5. Explain Hall effect.

6. Obtain an expression for mobility of current carriers (electrons).

7. Obtain an expression for Hall-voltage and Hall-coefficient.

(C) Long Answer Type Questions :

1. Discuss Sommerfield model of metal and hence derive the expression for energy of free electrons in metal.

2. Discuss the Fermi-Dirac distribution of electron in the metal.

3. Derive an expression for Fermi energy at absolute zero temperature.

4. Explain Hall effect. Obtain an expression for Hall voltage and Hall coefficient.

5. Distinguish between metals, semiconductors and insulators on the basis of band theory of metals.

6. Discuss the theory of origin of energy bands. Explain C.B., V.B. and forbidden energy gap in a crystal.

SOLID STATE DEVICE

SYLLABUS

Timer (IC 555), Block diagram, Function of each block, Pin configuration, Applications – Astable, Monostable, Bistable multivibrator.

5.1 INTRODUCTION

Timer is an electron circuit which generate waveform or pulse of definite time interval. IC-555 is an eight pin integrated circuit (IC) first introduced in 1971 by American company signetics. It has dual in line package (DIP) and has 25 transistors, 2 diodes and 15 resistors. The block diagram of IC-555 contains two comparators (op-amps), two transistors and a flip-flop (RS). This timer has its different versions. The improved versions are also available in 14 pin IC form, which are again dual-in-line packed (DIP).

IC-555 has number of applications in industry. Its timing pulse duration can vary from few milliseconds to some hours. Most common application of IC-555 is multivibrator. The accuracy and stability of output waveform of this timer is very high and hence it is widely used.

5.2 BLOCK DIAGRAM OF IC-555

It is the different electronic circuit present in IC-555. Fig. 5.1 shows the block diagram of IC-555.

The block diagram of IC 555 basically has two comparators (which are operational amplifiers), an R-S flip-flop, two transistors T_1, T_2; resistor network and an inverter (NOT gate). The resistive network is formed by three equal resistances R which form the voltage divider network. Due to this, voltage at a point (A) after lower resistance R is $\frac{1}{3}V_{CC}$ and that after the two lower resistances i.e. at point B is $\frac{2}{3}V_{CC}$. The V_{CC} is supply voltage which energises the circuit. Comparator I compares the threshold voltage with a reference voltage $+\frac{2}{3}V_{CC}$ volt. Similarly, comparator-II compares the trigger voltage with a reference voltage $\frac{1}{3}V_{CC}$ volt. The output of comparator I forms R input and output of comparator II forms S input of R-S flip-flop. The R-S flip-flop will take its state according to the output of two comparators. Transistor T_2 is discharge transistor whose collector is connected to the discharge terminal (pin-7) through lower end of resistive voltage divider network. This transistor T_2 saturates (turns on) or cut-off (off state) depending upon the output $\bar{Q}$ of R-S flip-flop (FF). The transistor T_1 provides reset input (pin-4) which sets the normal output Q to 0 and $\bar{Q}$ to 1 state; i.e., it resets the whole timer irrespective of any inputs.

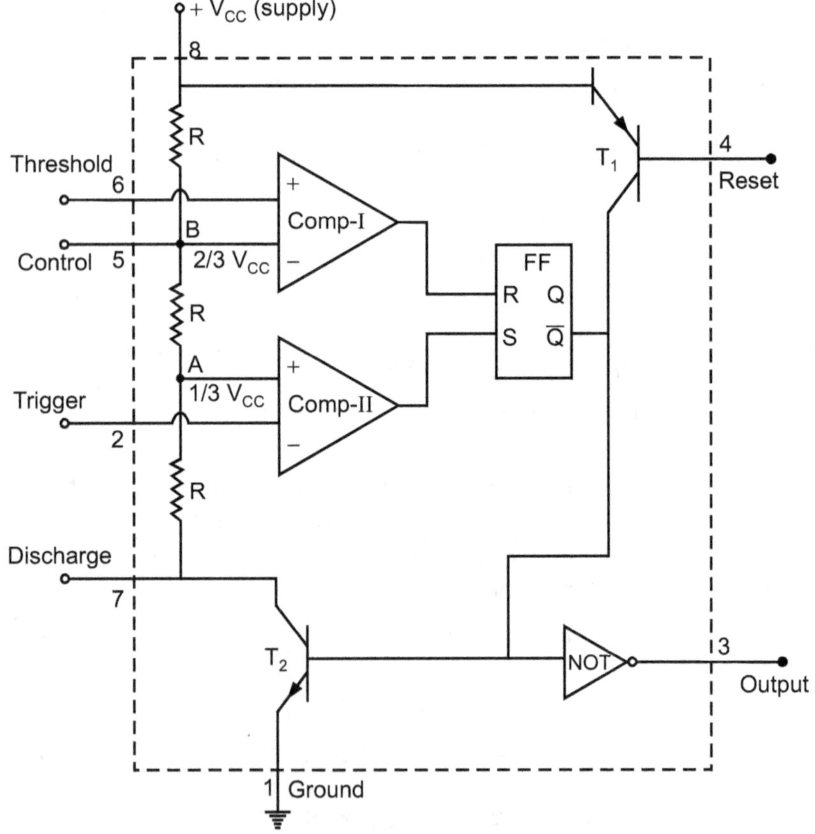

Fig. 5.1 : Block diagram of IC-555 timer

The function of each block mentioned above can be elaborated as follows.

Three internal resistances (R each) forming voltage divider arrangement divide the supply voltage $+V_{CC}$ such that it provides $\left(\frac{2}{3}\right) V_{CC}$ at inverting terminal of the comparator I and $\left(\frac{1}{3}\right) V_{CC}$ at the non-inverting terminal of comparator II. In most applications, the control input is not used and hence control voltage is the voltage at point B which is $\left(\frac{2}{3}\right) V_{CC}$.

Therefore comparator I has two inputs, threshold input (pin 6) and control input (pin 5). Whenever threshold voltage exceeds the control voltage, the output of comparator I will reset the flip-flop so that its output $\bar{Q}$ is high. This high output goes to two places – one at the input of NOT gate i.e. inverter and the other at the base of the transistor T_2. The output of flip-flop is inverted by NOT gate and forms the output of the timer (pin-3) which is low. Whereas, the high input to the base of transistor turns on the transistor T_2 so that the discharge of voltage at discharge terminal (pin 7) takes place through T_2 which provides low resistance path to the ground (pin 1). This condition remains until the comparator-II triggers the flip-flop. When the trigger voltage falls below $\left(\frac{1}{3}\right) V_{CC}$, then S input of flip-flop becomes high and output of the flip-flop

$\bar{Q}$ changes and becomes low (zero). This is the set state of the flip-flop.

This high $\bar{Q}$ output going to base of T_2 turns it off and also output of timer at pin 3 becomes high. These conditions will continue and comparator II gives output high which forces flip-flop output to high.

A voltage may be applied at control input (pin 5) to change the switching level of comparator I input from $\left(\frac{2}{3}\right) V_{CC}$ to some higher value. To obtain the switching, the voltage at threshold should exceed control voltage or $\left(\frac{2}{3}\right) V_{CC}$ whichever is higher. If control voltage is not applied then normally, a capacitor (of the order of nF) is connected between pin-5 and ground which avoids false triggering.

When reset input is low then irrespective of other inputs the output of flip-flop $\bar{Q}$ is high and hence the output of the timer goes low. When the reset is not used then it is connected to $+V_{CC}$. The working of IC-555 timer is summarised by considering following cases :

Case I : When no trigger threshold input is applied then comparator II output is high (1) and comparator I output is low (0). Hence $S = 1$ and $R = 0$. The flip-flop is in set state and $Q = 1$ and $\bar{Q} = 0$. Therefore output of timer at pin 3 is high (1).

Case II : When trigger pulse of negative voltage greater than $\left(\dfrac{1}{3}\right) V_{CC}$ is given, then $S = 0$ and hence $Q = 0$ and $\bar{Q} = 1$. Therefore output of timer is low.

Case III : When threshold voltage becomes higher than $\left(\dfrac{2}{3}\right) V_{CC}$ or control voltage, then $R = 1$ and hence $Q = 0$ and $\bar{Q} = 1$. Therefore output of the timer is again low.

Case IV : Irrespective of trigger and threshold voltage, when reset input is made low (0) then transistor T_1 turns on (no voltage drop across it) 1 and $+V_{CC}$ voltage appears at $\bar{Q}$. That is $\bar{Q}$ becomes 1, irrespective of R and S inputs and hence output of the timer is forced to low (0). Thus reset overrides trigger and trigger overrides threshold.

5.3 PIN CONFIGURATION OF IC-555

IC-555 is an integrated timer circuit which consists of different blocks as explained in Section 5.2. In nutshell, it is an eight pin integrated circuit (IC) dual-in-line packing as shown in Fig. 5.2. This IC designed in 1970s has different versions but most of them contains 25 transistors, two diodes and 15 resistors. They can be available in original bipolar form or low powered CMOS form. All these are available in 8-pin form as given in Fig. 5.2. Sometimes they are available in 14 pin form. These eight pins (four on each side) are explained as follows.

Pin-1 : This pin is giving ground terminal. All voltages are measured with respect to this ground terminal which is grounded (0 V).

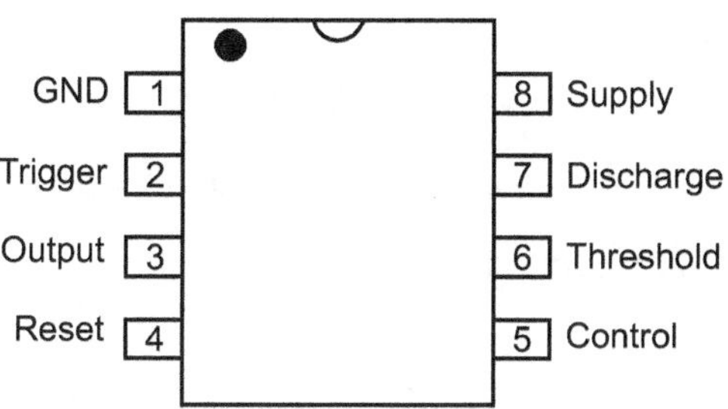

Fig. 5.2 : Pin configuration of IC-555

Pin-2 : It is the trigger terminal. This pin is input (inverting) to comparator II. It is responsible for transition of flip-flop from set to reset state as its output forms R input of flip-flop. A negative pulse of dc voltage level greater than $\left(\dfrac{1}{3}\right) V_{CC}$ when applied to this terminal then output of comparator II becomes high and hence output $\overline{Q}$ of flip-flop is forced to low and the output of timer becomes high.

Pin-3 : It is the output terminal of the IC 555 timer. The output of flip-flop $\overline{Q}$ is inverted and given at pin-3 as the timer output. This output is approximately driven to 1.7 V below V_{CC} or to ground.

Pin-4 : It is called as the reset terminal. It is used to reset or disable the timer. When the reset input is low then $\overline{Q}$ is high and Q is zero (or low). Hence the terminal is called reset terminal. When reset is not used then this terminal is connected to $+V_{CC}$ to avoid false triggering. Thus, reset input has direct control on the output of the timer irrespective of other inputs. Hence reset input overrides the trigger input which overrides the threshold voltage. From Fig. 5.1, when reset input is low (0) then transistor T_1 turns on, $\overline{Q}$ becomes high (1) and hence the output of the timer becomes low (0).

Pin-5 : It is called control voltage terminal. The switching voltage level of comparator I is decided by the voltage at this terminal and thus it changes the pulse width of the output. In short it provides controlled access to internal voltage divider.

Pin-6 : This is the threshold terminal. It forms the non-inverting input terminal of comparator I. When voltage at this terminal i.e. threshold voltage exceeds the $\left(\dfrac{2}{3}\right) V_{CC}$ or control voltage (whichever is higher) then output of comparator I goes low.

Pin-7 : It is called the discharge terminal. This pin is internally connected to collector of transistor (T_2) and a capacitor is connected between this terminal and ground. This terminal is called discharge terminal, because when transistor (T_2) turns on then capacitor discharges through this terminal and transistor. This terminal is also called as open collector output.

Pin-8 : It is called as supply voltage terminal. The voltage applied at this terminal $(+V_{CC})$ provides energy to the timer. A voltage from +5 V to +18 V can be applied between this terminal and ground (pin-1) depending on the version.

5.4 APPLICATIONS OF IC-555

This IC is a timer IC and has hundreds of applications depending upon the requirement of the circuit. However, the common application of it is the multivibrator. Multivibrators are nothing but the square wave oscillators or relaxation oscillators. These are two state devices and depending upon the stability in these states, the devices are classified into monostable, bistable and astable multivibrators.

5.4.1 Astable Multivibrator

These are electronic circuits which have no stable state. They are also called as free running multivibrators. The output waveform of this multivibrator is square wave whose period and frequency are determined by circuit components connected externally to IC 555. The circuit diagram of astable multivibrator using IC-555 is shown in Fig. 5.3.

Fig. 5.3 shows resistance R_1 connected between $+V_{CC}$ and discharge terminal (pin 7). The threshold (pin 6) and trigger (pin 2) terminals are connected together and then connected between resistance R_2 and capacitor C. The other end of the capacitor C is connected to the ground. The resistance R_2 is connected between resistance R_1 and capacitor C. The control terminal (pin 5) is connected to 10 nF capacitor whose other end is grounded. Terminal (pin 1) 1 is grounded and the reset (pin 4) and supply (pin 8) are connected to +V_{CC}. The output is taken at pin 3.

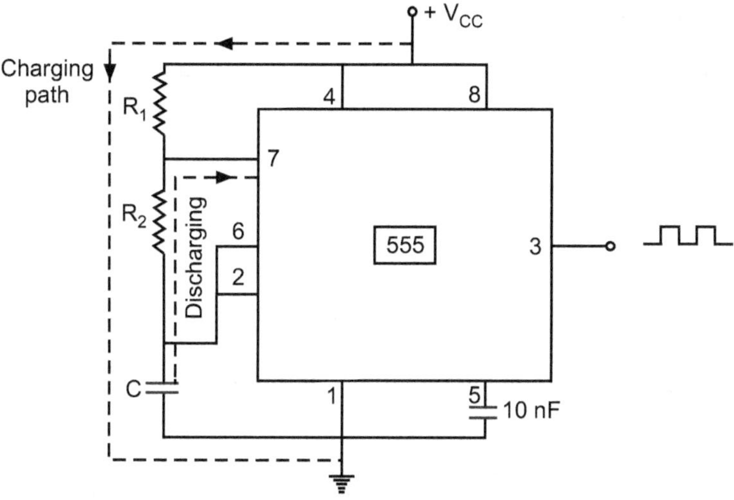

Fig. 5.3 : Circuit diagram of Astable Multivibrator

Working : The working of IC-555 timer as astable multivibrator can be explained as follows.

Let when the power is turned on the flip-flop is initially cleared and output of the inverter is high. The discharge transistor T_2 is off and capacitor C starts charging through R_1 and R_2. When capacitor voltage exceeds $\left(\dfrac{2}{3}\right) V_{CC}$ the comparator I output becomes high and it resets the flip-flop. Hence the $\bar{Q}$ output of flip-flop is high and output of timer is low. At the same time discharge transistor T_2 turns on and capacitor C starts discharging through R_2. Due to this, voltage across capacitor C and hence at terminal (pin) 2 (trigger) decreases. When this voltage becomes less than $\left(\dfrac{1}{3}\right) V_{CC}$ then comparator II output will be high and hence flip-flop gets set with Q = 1 and $\bar{Q}$ = 0. Therefore, final timer output becomes 1.

Now, the discharge transistor is off (as $\bar{Q} = 0$) and hence capacitor starts again through R_1 and R_2 as explained above. This process of charging and discharging of capacitor continues respectively so that output of square wave is generated.

Fig. 5.4 shows output waveform and capacitor voltage waveform. The output is square wave with time period T and T_{ON} represents the time for which the output of timer is high whereas T_{OFF} is the time for which output of the timer is low. Therefore, $T = T_{ON} + T_{OFF}$ and the ratio of T_{ON} and T is called duty cycle of astable multivibrator.

The capacitor voltage waveform is sawtooth waveform and the voltage varies between $\left(\dfrac{1}{3}\right) V_{CC}$ and $\left(\dfrac{2}{3}\right) V_{CC}$. The period and frequency of this waveform is same as that of output waveform.

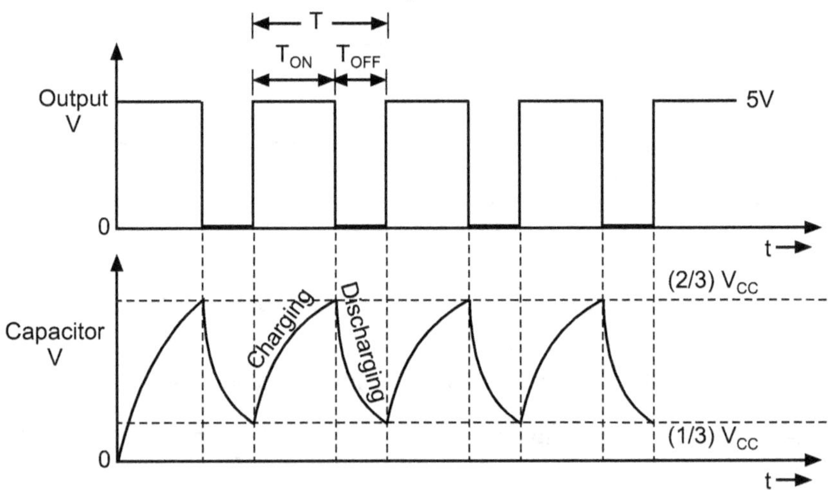

Fig. 5.4 : Output and capacitor voltage waveforms of Astable Multivibrator using IC 555

Period and Frequency of the wave :

As explained above capacitor charges through R_1 and R_2. Therefore, the charging time is given as

$$T_{charging} = T_{ON} = \log (2) \cdot (R_1 + R_2) \, C$$

$$\therefore \quad T_{ON} = 0.69 \, (R_1 + R_2) \, C \qquad \dots (5.1)$$

Similarly, the capacitor C discharges through R_2. Hence, discharge time is written as

$$T_{discharge} = T_{OFF} = 0.69 \, (R_2) \, C \qquad \ldots (5.2)$$

Hence, period of the wave is

$$T = T_{ON} + T_{OFF} = 0.69 \, (R_1 + 2R_2) \, C \qquad \ldots (5.3)$$

and $\qquad$ Frequency $= f = \dfrac{1}{T}$

$$\therefore \qquad f = \dfrac{1}{0.69 \, (R_1 + 2R_2) \, C} \qquad \ldots (5.4)$$

Duty cycle : It is the ratio of T_{ON} and T written as

$$\text{Duty cycle (D)} = \frac{T_{ON}}{T} = \frac{0.69 \, (R_1 + R_2) \, C}{0.69 \, (R_1 + 2R_2) \, C}$$

$$\therefore \qquad D = \frac{R_1 + R_2}{R_1 + 2R_2} \qquad \ldots (5.5)$$

and percentage duty cycle is

$$D\% = \frac{R_1 + R_2}{R_1 + 2R_2} \times 100\% \qquad \ldots (5.6)$$

From equation (5.6), when $R_2 \rightarrow 0$, then duty cycle becomes 100% or 1. But if $R_1 \rightarrow 0$ then duty cycle is 50% or 0.5. Thus duty cycle varies from 0.5 to 1. However, decrease the duty cycle below 0.5, which can be possible, a diode can be placed in parallel with R_2 which bypasses R_2 and T_{ON} depends on R_1 and C. Thus T_{ON} can be made less than T_{OFF} and hence duty cycle will be less than 0.5.

5.4.2 Monostable Multivibrator

Monostable multivibrator is an electronic circuit which has, as name suggests, one stable state. The circuit goes to the other unstable (quasi-stable) state when it is triggered, remains there for definite interval and then returns back on its own to the stable state. This monostable multivibrator can be designed using IC-555. Fig. 5.5 shows monostable multivibrator mode of operation of IC-555.

The circuit details in this Fig. 5.5 show that a resistance R is connected between pin 4, pin 8 and pin 7, pin 6. The capacitor C is connected between pin 6, pin 7 and ground. Trigger input pulse of voltage is given externally to pin 2. Pin 1 is grounded and pin 5 is grounded through C_1. The output of the circuit is taken at pin 3. The pin 4 and 8 are connected to $+V_{CC}$ which can be between 5 V to 15 V.

Working : Initially, the output of the circuit at pin 3 is low which is stable state of the circuit. The transistor T_2 (block diagram Fig. 5.1) is on and capacitor C is shorted to ground.

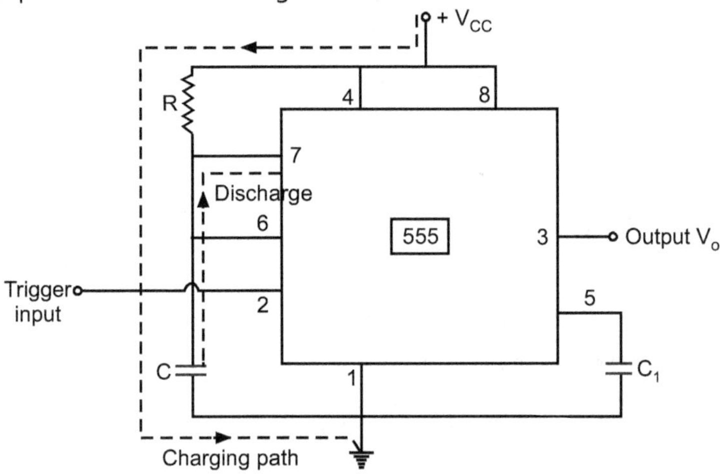

Fig. 5.5 : Monostable Multivibrator using IC-555

When a negative voltage pulse is applied at pin 2, the trigger input falls below $\left(\dfrac{1}{3}\right) V_{CC}$ and the output of comparator II goes high. This sets the flip-flop so that $Q = 1$ and $\bar{Q} = 0$. Hence, (i) transistor T_2 turns off, (ii) output of timer becomes high (1) and (iii) capacitor starts charging towards $+V_{CC}$ through resistance R with time constant RC. When voltage across the capacitor C increases and slightly becomes higher than $\left(\dfrac{2}{3}\right) V_{CC}$, then output of comparator II goes high (1). Hence, flip-flop resets and $Q = 0$ and $\bar{Q} = 1$. Hence (i) transistor T_2 turns on, (ii) output of the timer goes low (0), (iii) capacitor starts discharging through transistor T_2 (pin 7) almost instantly and (iv) the output returns back from quasi-stable state [output = high (1)] to stable state [output = low (0)]. This output of the circuit remains low until a next trigger pulse is applied.

When again trigger pulse is applied then above explained cycle of operation gets repeated. Thus for every trigger pulse, we get output of the multivibrator high for specific time T_{ON}.

The trigger voltage, output voltage and capacitor voltage waveforms are shown in Fig. 5.6.

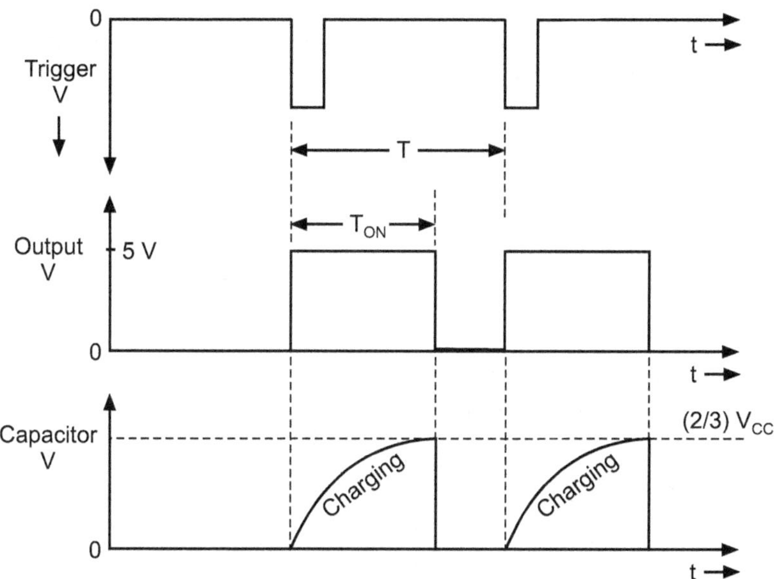

Fig. 5.6 : Trigger voltage, output voltage and capacitor voltage waveforms in Monostable Multivibrator using IC-555

Pulse width : Pulse width depends on the charging time constant RC. In other words, this time constant of the width of the output pulse i.e. the time for which the output is high (T_{ON}). The capacitor voltage V_C is

$$V_C = V_{CC} (1 - e^{-t/RC}) \qquad \qquad \dots (5.7)$$

The capacitor charges upto $\left(\dfrac{2}{3}\right) V_{CC}$. Hence above equation becomes

$$\frac{2}{3} V_{CC} = V_{CC} (1 - e^{-t/RC})$$

$$\therefore \qquad \frac{2}{3} = 1 - e^{-t/RC}$$

$$\therefore \qquad e^{-t/RC} = 1 - \frac{2}{3} = \frac{1}{3}$$

$$\therefore \qquad 3 = e^{t/RC}$$

$$\therefore \qquad \log 3 = \frac{t}{RC}$$

$$\therefore \qquad t = RC \log 3$$

$$\therefore \qquad \text{Pulse width} = t = T_{ON} = \log 3 \times RC$$

$$\therefore \qquad T_{ON} = 1.1 \times RC \qquad \qquad \dots (5.8)$$

This equation (5.8) gives pulse width in monostable multivibrator.

5.4.3 Bistable Multivibrator using IC 555

It is a multivibrator having two stable states. It remains in any one of the stable states for indefinite time unless it is triggered to the other stable state. Hence, it works as a memory cell or flip-flop and stores one bit of information.

The circuit diagram for bistable multivibrator using IC-555 is given in Fig. 5.7.

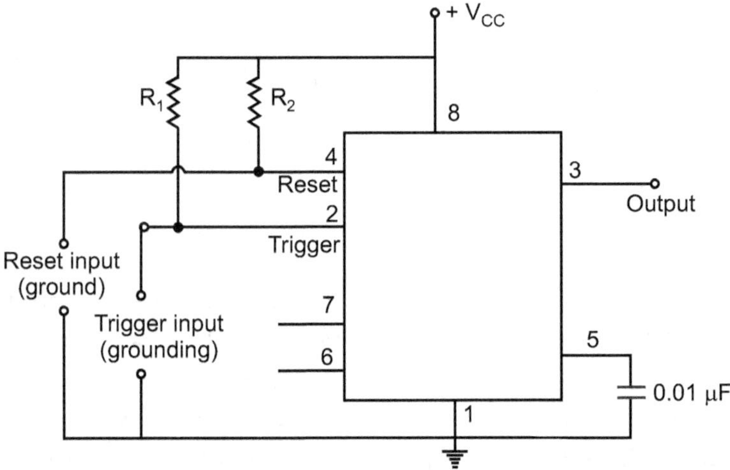

Fig. 5.7 : Circuit diagram of Bistable Multivibrator using IC-555

It is also called as Schmitt trigger. In this circuit the trigger input (pin 2) and reset input (pin 4) are held high through pull-up resistors R_1 and R_2 respectively. A momentary reset input and trigger input can be given to these terminals. The discharge (pin-7) and threshold (pin-6) terminals are kept floating. Pin 5 is connected to ground through a small value capacitor (0.01 to 0.1 μF). No timing capacitors are used in bistable multivibrator. Further, supply +V_{CC} is given to pin-8, output is taken at pin-3 and pin-1 is grounded.

Working : The working of this circuit can be explained as follows. In this circuit, generation of high and low outputs is not dependent on charging and discharging of capacitor in the RC unit but rather it is controlled by external trigger and reset input signals.

Initially, the output of the circuit is low. When the momentary trigger input is applied or pulled to ground then comparator II gives high output which sets the internal flip-flop. Hence, $\bar{Q}$ becomes zero and output of the circuit becomes high. This high output of the circuit remains until the momentary reset signal is made zero.

When reset input is pulled to the ground then it resets the internal flip-flop and hence output of the circuit becomes zero. Thus output of the circuit remains in high state from the moment of application of trigger (pulled to ground) to application of reset (pulled to ground) signals. These waveforms of trigger, reset and output voltage are shown in Fig. 5.8.

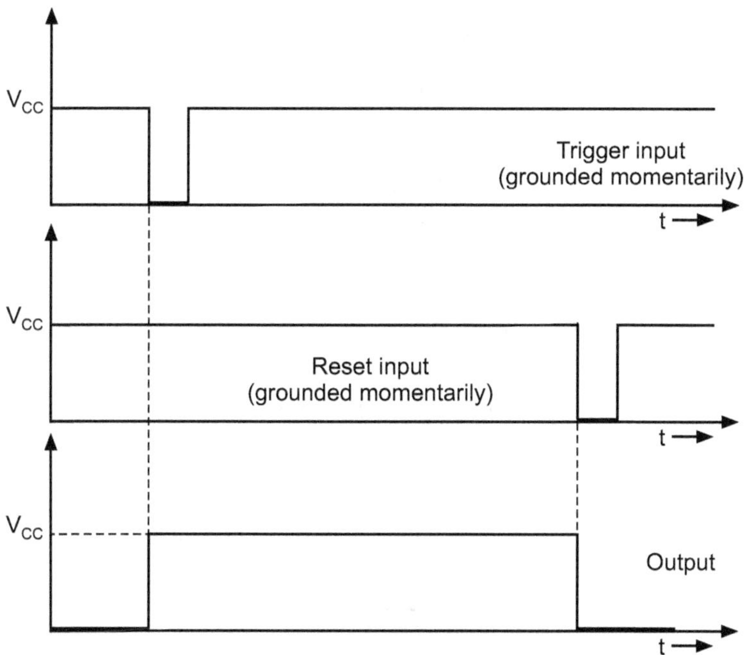

Fig. 5.8 : Trigger, reset and output waveforms in Bistable Multivibrator

The low output of the circuit obtained by grounding reset input remains low until the next trigger input is given. Thus in both the states, output is high and output is low, the circuit is stable. Hence it is called as bistable multivibrator.

SOLVED PROBLEMS

Problem 5.1 : In astable multivibrators formed by using IC-555, R_1 = 10 kΩ, R_2 = 2 kΩ and C = 0.1 μF. Calculate (i) frequency of the multivibrator, (ii) duty cycle.

Solution : (i) Frequency of the multivibrator is

$$f = \frac{1}{0.69 \, (R_1 + 2R_2) \cdot C}$$

$$\therefore \quad f = \frac{1}{0.69 \, (10 \times 10^3 + 2(2) \times 10^3) \times 0.1 \times 10^{-6}}$$

$$\therefore \quad f = \frac{1}{0.69 \, (14 \times 10^3) \times 0.1 \times 10^{-6}}$$

$$\therefore \quad f = \frac{1}{0.69 \times 14 \times 10^3 \times 10^{-7}}$$

$$\therefore \quad f = \frac{10^4}{0.69 \times 14}$$

$$\therefore \quad f = \frac{10^4}{9.66} = \textbf{1035.19 Hz}$$

(ii) Duty cycle is, $\quad D = \dfrac{T_{ON}}{T}$

But $\quad T_{ON} = 0.69 \, (R_1 + R_2) \, C$

and $\quad T = 0.69 \, (R_1 + 2R_2) \, C$

$$\therefore \quad D = \frac{R_1 + R_2}{R_1 + 2R_2} = \frac{10 \times 10^3 + 2 \times 10^3}{10 \times 10^3 + 2 \times 2 \times 10^3}$$

$$\therefore \quad D = \frac{12 \times 10^3}{14 \times 10^3}$$

$$\therefore \quad D = \frac{12}{14} = \textbf{0.857}$$

Problem 5.2 : In monostable multivibrator, 0.01 µF capacitor charges through a resistance of 5 kΩ. Find the pulse width of the output.

Solution : Pulse width t = T_{ON} = 1.1 RC

$$\therefore \qquad T_{ON} = 1.1 \times 5 \times 10^3 \times 0.01 \times 10^{-6}$$

$$\therefore \qquad T_{ON} = 1.1 \times 5 \times 10^3 \times 10^{-8}$$

$$\therefore \qquad T_{ON} = 5.5 \times 10^{-5} = 55.0 \times 10^{-6} \text{ sec}$$

$$\therefore \qquad T_{ON} = \mathbf{55 \text{ ms}}$$

EXERCISES

(A) Multiple Choice Type Questions :

1. Number of comparators in IC 555 are ......

    (a) two   (b) three

    (c) four   (d) five

2. How many R-S flip-flops are present in IC-555 ?

    (a) four   (b) three

    (c) two   **(d) one**

3. Which of the following is not a terminal of IC-555 ?

    (a) threshold   (b) trigger

    (c) gate   (d) reset

4. In IC 555, which terminal is connected to the collector of the transistor?

    (a) control   (b) trigger

    (c) discharge   **(d) threshold**

5. Basic version of IC-555 has ...... pins.

    (a) 8   (b) 10

    (c) 12   (d) 14

6. If control terminal voltage is not given externally then it is ...... volt.

    (a) V_{CC} (b) $\dfrac{V_{CC}}{3}$

    (c) $\dfrac{2V_{CC}}{3}$ (d) $\dfrac{V_{CC}}{2}$

7. Voltage at comparator II input terminal is ......

   (a) V_{CC}

   (b) $\dfrac{V_{CC}}{3}$

   (c) $\dfrac{2V_{CC}}{3}$

   (d) $\dfrac{V_{CC}}{2}$

8. When timer is reset then its output is ......

   (a) zero

   (b) high

   (c) V_{CC}

   (d) $2V_{CC}$

9. Astable multivibrator has ...... stable state/states.

   (a) 3

   (b) 2

   (c) 1

   (d) 0

10. Monostable multivibrator has ...... stable state/states.

    (a) 0

    (b) 1

    (c) 2

    (d) 3

11. Number of stable states in bistable multivibrator is ......

    (a) 0

    (b) 1

    (c) 2

    (d) 1 or 2

12. The output state of IC-555 timer cannot be changed by changing voltage at ...... terminal.

    (a) reset

    (b) trigger

    (c) threshold

    (d) discharge

13. Pulse width in monostable multivibrator is directly proportional to ......

    (a) RC

    (b) $(R_1 + R_2)\, R_C$

    (c) $(R_1 + 2R_2)\, C$

    (d) $(2R_1 + R_2)\, C$

14. Frequency of astable multivibrator depends on ......

    (a) only R_1　　　　　　　　(b) only R_2

    (c) only C　　　　　　　　 **(d) R_1, R_2 and C**

15. Duty cycle in astable multivibrator varies between ......

    (a) 0 to 0.5　　　　　　　**(b) 0.5 to 1**

    (c) 1 to 1.5　　　　　　　　(d) 0 to 1.5

16. Which of the following multivibrators is used as a memory cell ?

    (a) astable　　　　　　　　(b) monostable

    (c) bistable　　　　　　　(d) unstable

17. Which multivibrator requires no trigger input ?

    (a) quasi-stable　　　　　　**(b) astable**

    (c) monostable　　　　　　　(d) bistable

(B) Short Answer Type Questions :

    1.   Draw a block diagram of IC-555.

    2.   Give pin configuration of IC-555.

    3.   Explain the meaning of each terminal.

    4.   Sketch the monostable multivibrator mode of operation of IC-555.

    5.   Draw a circuit diagram for astable multivibrator using IC-555.

    6.   Draw a circuit using IC-555 to work as bistable multivibrator.

    7.   Derive expression for pulse width of monostable multivibrator.

    8.   Sketch the waveforms in monostable multivibrator.

(C) Long Answer Type Questions :

    1.   Draw a block diagram of IC-555 and explain the function of each block.

2. Give the pin configuration of IC-555 and explain the meaning of each terminal.

3. Draw a circuit diagram for astable multivibrator using IC 555 and explain its working. Find the expressions for frequency and duty cycle of this multivibrator.

4. Sketch the circuit diagram for monostable multivibrator using IC 555 and explain its working. Derive expression for its pulse width.

5. Draw a circuit diagram for Bistable multivibrator using IC 555. Explain its working. Sketch the different waveforms of this multivibrator.